Optimizing Your Modernization Journey with AWS

Best practices for transforming your applications and infrastructure on the cloud

Mridula Grandhi

BIRMINGHAM—MUMBAI

Optimizing Your Modernization Journey with AWS

Associate Group Product Manager: Preet Ahuja

Publishing Product Manager: Surbhi Suman

Senior Editor: Divya Vijayan

Technical Editor: Rajat Sharma

Copy Editor: Safis Editing

Project Coordinator: Ashwin Kharwa

Proofreader: Safis Editing

Indexer: Rekha Nair

Production Designer: Jyoti Chauhan

Marketing Coordinator: Rohan Dobhal

First published: July 2023

Production reference: 1070623

Published by Packt Publishing Ltd.
Livery Place
35 Livery Street
Birmingham
B3 2PB, UK.

ISBN 978-1-80323-454-0

www.packtpub.com

To all those who inspire, influence, educate, motivate, and lead by example.

– Mridula Grandhi

Contributors

About the author

Mridula Grandhi is a senior leader of solutions architecture specializing in the **Amazon Web Services** (**AWS**) Compute portfolio of services such as containers, serverless, Graviton, and hybrid services. She has more than 16 years of experience architecting and building distributed software systems across industry verticals such as the supply chain, the automotive industry, telecommunications, and financial services. In her current leadership position, she works with AWS customers and provides strategic guidance on optimal pathways to modernize their workloads and achieve their business objectives.

I want to thank all those in my life who see me for who I am, celebrate every dimension of who I am, and give me permission to be the way I am.

About the reviewers

Mario Mercado has worked in software development for around 10 years in different roles, including developer, **quality assurance** (**QA**) engineer, DevOps engineer, and even director of cloud infrastructure. Mario has helped multiple companies in the last few years succeed on their cloud journeys while keeping cost optimization and automation at the forefront. In addition, Mario has worked with the community, sharing his knowledge on social media and on other platforms such as the AWS Community Builders program and working with the teaching platform, A Cloud Guru, to create courses about Amazon **Elastic Kubernetes Service** (**EKS**) and AWS CodeStar, for example. His result-oriented focus has been his most important tool, as he knows what businesses need and how it translates to topics in his area of expertise.

I am truly honored to have worked with Packt on this book. I always thought about that while reviewing each chapter, which I highly enjoyed. It was a huge pleasure for me to work with this amazing team, and I especially acknowledge the outstanding work of the author, Mrindula. Thanks to all the Packt team members who coordinated my contribution to this book!

Vikas Kanwar is a solution strategist for cloud services and technology adoption who aims to create and capture value for clients. Vikas has 15 years of experience in IT and has worked with large organizations and global enterprises such as AWS, Wipro Ltd., Gartner Inc., and HCL. Vikas has been responsible for cloud pre-sales and delivery, architecture and design, cloud transformation, and data center migration for large clients across geographies. Vikas previously published papers on cloud computing while working as an analyst with Gartner.

I truly believe that technological advancement has changed our lives and that with the myriad of choices made available by cloud computing, it has become easier for people with a solutions-oriented mindset to innovate and make a positive impact. Thank you to all these people who make this area of technology exciting and rewarding.

Table of Contents

2

Understanding Cloud Migration   35

3

Preparing for Cloud Migration   51

4

Implementing Cloud Migration Strategies   73

Part 2: Cloud Modernization – Application, Data, Analytics, and IT

5

Modernization in the Cloud   97

6

Application Modernization Approaches   133

7

Application Modernization – Compute   169

8

Implementing Compute and Integration on the Cloud Using AWS   215

Part 3: Security and Networking Transformation

11

Transforming Networking on the Cloud Using AWS   309

Part 4: Cloud Economics, Compliance, and Governance

12

Operating on the Cloud with AWS   335

Preface

Modernization requires a multi-dimensional approach. There is a lot to understand when you embark on a digital transformation journey and modernize your platforms, which includes building new applications and retiring legacy solutions. As your journey to the cloud matures, you will want to accelerate your modernization and maximize its value. This book is designed to help you develop appropriate knowledge on the various aspects of modernization using **Amazon Web Services** (**AWS**). You will gain insights that will enable you to make decisions and reach new levels of operational efficiency, increased scalability, and improved performance and resiliency on the cloud.

As per the Gartner research as of 2022, 70% of workloads will be on the cloud by 2024, and organizations need a cloud strategy to leverage the advantages of cloud computing. A cloud strategy is a living document that guides teams to leverage the advantages of cloud computing within their organization. Many organizations have no cloud strategy or only think they have one.

A cloud strategy is not an executive mandate or the same as an implementation/adoption/migration plan. Cloud-first is a common principle specified in cloud strategies. By itself, it is not a strategy. Devising a broad strategy requires the alignment of multiple stakeholders and accommodating scenarios, cloud services, vendors, and non-cloud environments.

This book covers a phased approach when it comes to relevant topics on modernization, not just for applications but also for databases, storage, networks, and security. By referencing the best practices and adoption patterns, this book provides a comprehensive view of the knowledge required to consume new technology and to deliver portfolio, application, and infrastructure value faster. This book will also guide you through strategies to simplify your business operations, architecture, and overall engineering practices.

We begin with introducing cloud migration for your existing workloads, evaluating modernization readiness for your workloads, prescriptive guidance for cloud migration, and choosing a cloud vendor. This is then followed by a modernization roadmap for your workloads with AWS services for technology categories such as Compute, Storage, Databases, Security, Networking, and Cloud Operations. Each chapter includes use cases for using the AWS services and case studies on how companies of various industries can take advantage of modernization.

Who this book is for

IT and business executives, program and project managers, product owners, and operations and infrastructure owners can gain insights into how to strategize and implement modernization pathways for their workloads on the cloud with AWS using this book.

The main personas who are the target audience of this content are as follows:

Business and technology executives: Whether you are the leader of a start-up or an enterprise, you are the decision-maker and need to be informed on what makes the best business sense. This book will guide you through the cloud options to give you a broad range of insights and help you navigate through the core aspect of cloud modernization.

Architects: Architects need to have a strong background in cloud computing and an understanding of the breadth of cloud computing services and processes. This book guides you through the design principles and architectural patterns so you are able to bridge the gaps between complex business problems and solutions in the cloud.

Engineers and administrators: You will work with architects to ensure that the right technology or technologies are being built. This role has a wide impact and provides deep value to make sure the facets of the **Software Development Life Cycle (SDLC)** are being continuously improved using a mix of practices, tools, and technologies. This book will guide you through the concepts of cloud modernization functions and help you grow in your career to be a lead, manager, and so on.

What this book covers

Chapter 1, *Introduction to Cloud Transformation*, establishes a foundation by introducing the cloud and its key characteristics. It will also cover the motivators for cloud adoption and the different cloud service providers that are available on the market. It shares the details on cloud service models such as IaaS, PaaS, and SaaS, and explores the different deployment models, such as private, public, hybrid, multi-, and community clouds.

Chapter 2, *Understanding Cloud Migration*, provides an overview of the key concepts and fundamentals of cloud migration to support the understanding of the AWS pillars, such as operational excellence, security, reliability, performance efficiency, and cost optimization. The content covers the common cloud migration challenges and strategies to navigate through those challenges. It also builds on the cloud-first mindset and characterizes various phases of cloud migration.

Chapter 3, *Preparing for Cloud Migration*, reviews the basics of cloud migration and provides an overview of the common levers that drive cloud migration for any organization. It also provides a baseline understanding of concepts such as CapEx and OpEx and guidance on choosing the right cloud partner for your business. You will learn about a multi-cloud strategy and the best practices to prepare your workloads to run successfully on the cloud. Subsequent chapters build on the information provided in this chapter.

Chapter 4, *Implementing Cloud Migration Strategies*, begins with an introduction to cloud migration strategies and explains how to build a business case for cloud migration. It then dives deep into the six Rs of migration on the cloud and some best practices on how to achieve success through migration efforts. Finally, you will learn about the AWS services that you can leverage to accelerate the migration of your workloads on the AWS cloud.

Chapter 5, Understanding Modernization on Cloud, covers the aspects involved in modernizing your workloads on the cloud, and considerations when doing so. It then discusses the path to modernization on the cloud and the various guidelines involved at each step of the path. We go into detail on the benefits of modernization and how to get started with modernization on AWS. The chapter also covers the fundamental technology categories so that the subsequent deep dives in the next chapters have a baseline.

Chapter 6, Understanding Application Modernization Approaches, begins with an introduction to application modernization. It then focuses on the key strategies for you to consider while implementing application modernization. You will learn about the design principles to transform monoliths into microservices, and the best practices to align with. The chapter will showcase the various AWS services to help with application modernization and will conclude with a case study.

Chapter 7, Executing Compute Transformation on Cloud Using AWS, provides an overview of the compute services available on AWS, use cases and functionalities, and how they are implemented in AWS using a case study. It dives deep into container technologies and their implementation on AWS and provides case studies on containers and serverless.

Chapter 8, Implementing Compute and Integration on Cloud Using AWS, expands on the application modernization services through serverless and integration services. Case studies for these will be covered and you will learn how to implement these services on AWS. This chapter discusses the AWS offerings using which you can navigate on your application modernization journey effectively.

Chapter 9, Transforming Data and Analytics on Cloud Using AWS, discusses data on the cloud. It begins with an overview of data modernization, then dives deep into strategies for data modernization, and services to use on AWS that can be leveraged to process, store, and manage data on the cloud. It also reviews analytics on the cloud, and various AWS services to achieve optimal results of modernizing analytics on the cloud.

Chapter 10, Enabling Security on Cloud Using AWS, describes why security is job zero and highlights the importance of having a strong security posture on the cloud. You will learn about the spectrum of security options on AWS, and their use cases, features, functionalities, and case studies.

Chapter 11, Implementing Networking on Cloud Using AWS, provides you with an understanding of the networking concepts on the cloud and next-generation networking requirements. We will go into detail on the various strategies to operate a network on the cloud. AWS Networking services will be discussed, along with case studies and example use cases where network performance is important for applications.

Chapter 12, Achieving Operational Excellence on Cloud Using AWS, covers concepts related to cloud operations and what you need to build a successful cloud center of excellence. We go into detail on the various pillars of the AWS Well-Architected Framework to achieve high availability, scalability, optimized cost usage, sustainability, and reliability. You will learn about cloud financial management and site reliability engineering and how to use AWS services for the SRE practice.

Chapter 13, *Wrapping Up and Looking Ahead*, sums up all the important aspects that we discussed in the past 12 chapters. It will discuss the various design elements, architectures, latest technologies, and trends to look for.

To get the most out of this book

You will need to have an understanding of the basics of the following:

- Software architecture fundamentals

- Good understanding of the cloud

- Fundamentals of **AWS**

- Understanding of software engineering and the **SDLC**

Download the color images

We also provide a PDF file that has color images of the screenshots and diagrams used in this book. You can download it here: `https://packt.link/V0joL`.

Conventions used

There are a number of text conventions used throughout this book.

`Code in text`: Indicates code words in text, database table names, folder names, filenames, file extensions, pathnames, dummy URLs, user input, and Twitter handles. Here is an example: "You can define keys such as `customer_id or transaction_id` and deliver them to Amazon S3.

A block of code is set as follows:

```
exports.handler = async (event) => {
  return {
    statusCode: 200,
    body: JSON.stringify({ msg: "Hello from Lambda!" })
  };
};
```

Bold: Indicates a new term, an important word, or words that you see onscreen. For instance, words in menus or dialog boxes appear in **bold**. Here is an example: "Using the multi-tenant model, cloud **resources** are **pooled** via resource pooling."

> Tips or Important Notes
> Appear like this.

Get in touch

Feedback from our readers is always welcome.

General feedback: If you have questions about any aspect of this book, email us at customercare@packtpub.com and mention the book title in the subject of your message.

Errata: Although we have taken every care to ensure the accuracy of our content, mistakes do happen. If you have found a mistake in this book, we would be grateful if you would report this to us. Please visit www.packtpub.com/support/errata and fill in the form.

Piracy: If you come across any illegal copies of our works in any form on the internet, we would be grateful if you would provide us with the location address or website name. Please contact us at copyright@packtpub.com with a link to the material.

If you are interested in becoming an author: If there is a topic that you have expertise in and you are interested in either writing or contributing to a book, please visit authors.packtpub.com.

Share Your Thoughts

Once you've read *Optimizing Your Modernization Journey with AWS*, we'd love to hear your thoughts! Scan the QR code below to go straight to the Amazon review page for this book and share your feedback.

https://packt.link/r/1803234547

Your review is important to us and the tech community and will help us make sure we're delivering excellent quality content.

Download a free PDF copy of this book

Thanks for purchasing this book!

Do you like to read on the go but are unable to carry your print books everywhere?

Is your eBook purchase not compatible with the device of your choice?

Don't worry, now with every Packt book you get a DRM-free PDF version of that book at no cost.

Read anywhere, any place, on any device. Search, copy, and paste code from your favorite technical books directly into your application.

The perks don't stop there, you can get exclusive access to discounts, newsletters, and great free content in your inbox daily

Follow these simple steps to get the benefits:

1. Scan the QR code or visit the link below

https://packt.link/free-ebook/9781803234540

2. Submit your proof of purchase
3. That's it! We'll send your free PDF and other benefits to your email directly

Part 1:
Migrating to the Cloud

In this part, we will introduce cloud migration and why companies are moving some or all of their data center capabilities to the cloud. We will cover the benefits of migrating to the cloud, the common cloud migration challenges, and how to make a business case for cloud migration. We will take a deeper dive into why you need a plan that covers the technicalities behind cloud migration and also gets buy-in from your stakeholders. As we perform the cloud migration analysis, you will learn to evaluate your the current state of your business and IT environment and carry out business process mapping to improve the way your organization operates. You will learn about the various cloud migration strategies you can apply to fit your needs and achieve your business outcomes most efficiently.

This part comprises the following chapters:

- *Chapter 1, An Introduction to Cloud Transformation*
- *Chapter 2, Understanding Cloud Migration*
- *Chapter 3, Preparing for Cloud Migration*
- *Chapter 4, Cloud Migration Strategies*

Introduction to Cloud Transformation

1

Innovation, **efficiency**, and **profitability** are some of the main tenets for businesses to be able to adapt to the changing needs of the world. Amazon, Microsoft, and Apple are organizations that effectively continue to strive for innovation and manage to reinstate their successes at multiple turning points in their journeys. Netflix was able to reinvent its business and make the streaming experience more enjoyable by innovating its platforms so that they can withstand disruptions. Technology plays a crucial role in helping organizations increase their innovation and agility.

Cloud transformation becomes a topmost priority for organizations who want to explore, improve their day-to-day operations, and succeed in their businesses in these constantly changing times. Businesses around the world are embracing the cloud to supercharge their organization's growth, as well as innovating, running, scaling, delivering, optimizing, and mitigating any business risks quickly and efficiently. Cloud transformation often poses barriers that are difficult to break down and requires a clear vision of where to start.

In this chapter, we will cover the following topics:

- Introduction to the cloud
- Key characteristics of cloud computing
- Motivators for cloud adoption
- Cloud service providers at a glance – AWS, GCP, Azure, and more
- Service models (IaaS, PaaS, and SaaS)
- Exploring the deployment models (private, public, hybrid, multi, and community)

Introduction to the cloud

Many aspects of our everyday life have been transformed by ever-evolving digital solutions. Technology is changing rapidly, and industries are adapting to these changes at a rapid pace. **The cloud** has become the dominant term in technology in the past few years and the impact it's brought to businesses is beyond resounding. Before we learn more about cloud transformation, let's look at the cloud and what cloud computing is.

> **Cloud transformation**
>
> Cloud transformation is the step-by-step process of moving your workloads from local servers to the cloud. It is a process that brings technology and organizational processes together to accelerate the development, implementation, and delivery of new services.

The cloud is referred to as a collection of software, servers, storage, databases, networking, analytics, and intelligence that can be accessed via the internet instead of being locally available on your computer or device. These services are delivered through data centers located across the globe and linked through the internet.

The following diagram depicts the use of cloud computing and the accessibility of various devices via the cloud:

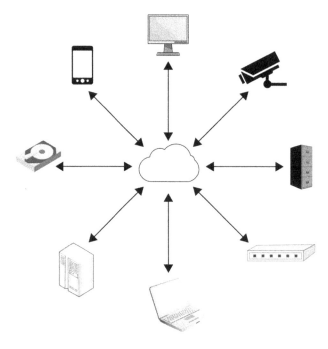

Figure 1.1 – Cloud computing

Before we dive deep into cloud computing, let's learn how we got here and how cloud computing began.

The origins of cloud computing

The history of cloud computing began almost 70 years ago, when corporations and large organizations began exploring computers and mainframe systems. In the 1950s and 1960s, these were only a reality for organizations with sufficient financial resources. Computers were simply large, expensive interfaces that required human operators to interact with the mainframe computer terminals to process complex data.

These early mainframe clients had limited computing power and needed the bulk of the available physical *servers* to get the work done. This model of computing is the predecessor of cloud computing.

In the 1970s, hardware-assisted virtualization was first introduced by IBM, which allowed organizations to run many virtual servers on a physical server at a given time. This was a milestone for mainframe owners as they could run virtual machines using the VM's operating system. Virtualization has come a long way and VM operating systems are a deployment option for many organizations for building and deploying applications. Today's cloud computing model couldn't have been possible without the concept of virtualization.

Technically, the concept of virtualization evolved with the internet as businesses started providing **virtual private networks** as a paid service. This resulted in a great momentum back in the 1990s leading to the development of a foundational block for modern cloud computing.

The term **cloud computing** specifies where the boundaries of computing follow the economic rationale rather than the technical limits alone.

> ### Virtualization
> Virtualization is the process of running a virtual instance by creating an abstraction layer over dedicated amounts of CPU, memory, and storage that are borrowed from a physical host computer.

In the following diagram, each VM runs an **operating system (OS)** of choice with its own software, libraries, and so on that are needed for its applications. This VM silo runs the hypervisor on top of the *bare-metal environment*:

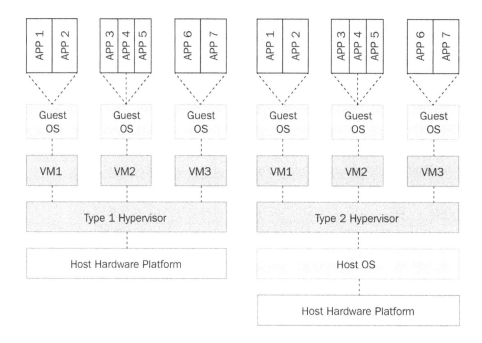

Figure 1.2 – Cloud computing

This virtualization technique forms the foundational component of cloud computing, where a **hypervisor** runs on a real machine and creates virtual operating systems on that particular machine.

In 2006, Amazon launched **Amazon Web Services** (**AWS**), the first cloud provider to offer online services to other websites of customers. In 2007, IBM, Google, and several other interested parties such as Carnegie Mellon University, MIT, Stanford University, the University of Maryland, and the University of California at Berkeley joined forces and developed research projects. Through these projects, they realized that computer experiments can be conducted faster and for cheaper by renting virtual computers rather than using their hardware, programs, and applications. The same year also saw the birth of Netflix's video streaming service, which uses the cloud and has revolutionized the practice of **binge-watching**.

> ### Cloud Computing
> AWS states that "*Cloud computing is the on-demand delivery of IT resources over the internet with pay-as-you-go pricing. Instead of buying, owning, and maintaining physical data centers and servers, you can access technology services, such as computing power, storage, and databases, on an as-needed basis from a cloud provider*".

This completes our introduction to the cloud, the history of cloud computing, and its evolution. In the next section, we will look at the key characteristics of cloud computing so to understand how cloud computing is beneficial for businesses in this new computing era.

Key characteristics of cloud computing

According to the **National Institute of Standards and Technology (NIST)**, cloud computing has six essential characteristics, as follows:

- On-demand self-service
- Wide range of network access
- Multi-tenant model and resource pooling
- Rapid elasticity
- PAYG model
- Measured service and reporting

We will discuss each of these in the following subsections.

On-demand self-service

In traditional enterprise IT settings, companies used to build the required infrastructure to run their applications locally; that is, **on-premises**. This means that the enterprises must set up the server's hardware, software licenses, integration capabilities, and IT employees to support and manage these infrastructure components. Because the software itself resides within an organization's premises, enterprises are responsible for the security of their data and vulnerability management, which entails training IT staff to be aware of security vulnerabilities and installing updates regularly and promptly.

Cloud computing is different from traditional IT hosting services since consumers don't have to own the required infrastructure to run their applications. With the cloud, a third-party provider will host and maintain all of this for you. Provisioning, configuring, and managing the infrastructure is automated in the cloud, which reduces the time to streamline activities and make decisions about capacity and performance in real time.

> Cloud automation
>
> Cloud automation is the process of automating tasks such as discovering, provisioning, configuring, scaling, deploying, monitoring, and backing up every component within the cloud infrastructure in real time. This involves streamlining tasks without human interaction and caters to the changing needs of your business.

On-demand makes it possible for consumers to benefit from the resources on the cloud *as and when required.* The cloud supplier caters to the demand in a real-time manner, enabling the consumers to decide on when and how much to subscribe for these resources. The consumers will have full control over this to help meet their evolving needs.

The self-service aspect allows customers to procure and access the services they want instantaneously. Cloud providers facilitate this via simple user portals to make this quick and easy. For example, a cloud consumer can request a new virtual machine and expect it to be provisioned and running within a few minutes. On-premises procurement of the same typically takes 90-120 days and also requires accurate forecasting to purchase the required RAM specifications and the associated hardware for a given business use case.

Wide range of network access

Global reach capability is an essential tenet that makes cloud computing accessible and convenient. Consumers can access cloud resources they need from anywhere, and from any device over the network through standard mechanisms such as authentication and authorization. The availability of such resources from thin or thick client platforms such as tablets, PCs, smartphones, netbooks, personal digital assistants, laptops, and more help the cloud touch every possible end user.

Multi-tenant model and resource pooling

Multi-tenancy is one of the foundational aspects that makes cloud services practical. To help you understand multitenancy, think of the safe-deposit boxes that are located in banks, which are used to store your valuable possessions and documents. These assets are stored in isolated and secure vaults, even though they're stored in the same location. Bank customers don't have access to other deposit boxes and are not even aware of or interact with each other. Bank customers rent these boxes throughout their lifetime and use security mechanisms to provide identification and access to their metal boxes. In cloud computing, the term multi-tenancy has a broader meaning, where a single instance of a piece of software runs on a server and serves multiple tenants.

> **Multi-tenancy**
> Multi-tenancy is a software architecture in which one or more instances of a piece of software are created and executed on a server that serves multiple, distinct tenants. It also refers to shared hosting, where server resources are divided and leveraged by end users.

The following diagram shows single-tenant versus multi-tenant models, both of which can be used to design software applications:

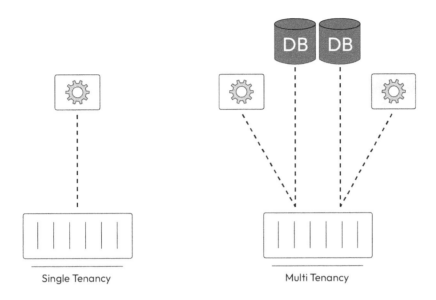

Figure 1.3 – Single-tenancy versus multi-tenancy

As an example of a multi-tenancy model, imagine an end user uploading content to social media application(s) from multiple devices.

Using the multi-tenant model, cloud resources are pooled via **resource pooling**. The intention behind resource pooling is that the consumers will be provided with ways to choose from an infinite pool of resources on demand. This creates a sense of immediate availability to those resources, without them being bound to any of the limitations of physical or virtual dependencies.

> **Resource pooling**
> Resource pooling is a strategy where cloud-based applications dynamically provision, scale, and control resource adjustments at the meta level.

Resource pooling can be used for services that support data, storage, computing, and many more processing technologies, thereby facilitating dynamic provisioning and scaling. This enables on-demand self-service for services where consumers can use these services and change the level of their usage as per their needs. Resource pooling, coupled with automation, replaces the following mechanisms:

- Traditional mechanisms
- Labor-intensive mechanisms

With new strategies that rely on increasingly powerful virtual networks and data handling technologies, cloud providers can provide an abstraction for resource administration, thereby enhancing the consumer experience of leveraging cloud resources.

Rapid elasticity

Elasticity is one of the most important factors and experts indicate this as the major selling point for businesses to migrate from their local data centers. End users can take advantage of seamless provisioning because of this setup in the cloud.

What is cloud elasticity? What are the benefits?

Before we answer these questions, let's take a look at the definition of elasticity.

Elasticity in the cloud refers to the end user's ability to acquire or release resources automatically to serve the varying needs of a cloud-based application while remaining operational.

Another criterion that is used in the cloud is scalability. Let's look at what it is and how it differs from cloud elasticity.

Scalability in the cloud refers to the ability to handle the changing needs of on-demand by either adding or removing resources within the infrastructure's boundaries.

Although the fundamental theme of these two concepts is adaptability, both of these differ in terms of their functions.

Scalability versus Elasticity

Scalability is a **strategic** resource allocation operation, whereas elasticity is a **tactical** resource allocation operation. Elasticity is a fundamental characteristic of cloud computing and involves taking advantage of the scalable nature of a specific system.

The inherent nature of dynamically adapting capacity helps businesses handle heavy workloads, as well as ensure that their operations go uninterrupted.

For example, take an online retail shipping website that is experiencing sudden bursts of popularity and their volume of transactions is peaking. To handle the workload, the website can leverage the cloud's rapid elasticity by adding resources to meet the transaction spikes. When the workloads do not have to meet such peaks, the services can be taken down just as quickly as they were added. You only pay for the services that you use at any given point.

Automatically commissioning and decommissioning resources is inherent to cloud elasticity and can be used to meet the in and out demands of businesses, thereby helping them manage and maintain their **operating expenditure** (**OpEx**) costs without having to put in any upfront **capital expenditure** (**CapEx**) costs and being locked into any long-term contracts.

PAYG model

The pay-per-use or **Pay As You Go** (**PAYG**) pricing model is a major highlight that's geared toward an economic model for organizations and end users. The per-second billing pricing plans that are provided by the cloud providers make it easy for businesses to witness a major shift from CapEx to OpEx. This enables the businesses to not worry about the upfront capital that they need to spend on on-premises infrastructures and capacity planning to meet ongoing demands. The traditional self-provisioning processes are often prone to extreme inefficiency and waste due to the complex supply chain model, which usually involves seamless communication between decision-makers and stakeholders.

However, cloud-based architectures and their inherent design models allow you to scale up your applications on the cloud during peak traffic and scale back down during periods where they're not needed as much, without having to worry about annual contracts or long-term license termination fees.

What are CapEx and OpEx?

CapEx involves funds that have been incurred by businesses to acquire and upgrade a company's fixed assets. This includes expenditures toward setting up the technology, the required hardware and software to run the services, and more.

OpEx involves the expenses that have been incurred by businesses through the course of their normal business operations. Such expenses include property maintenance, inventory costs, funds allocated for research and development, and more.

The businesses witness heavy OpExs when it comes to service and software procurement and management, tasks that are often expensive and inefficient. This model also often leads to complex payment structures and makes it difficult for businesses to fluctuate their usage. With the **PAYG** model, you pay for the resource charges for user-based services, versus an entire infrastructure. Once you stop using the service, there is typically no fee to terminate, and the billing for that service stops immediately.

Let's look at an example of how the PAYG model is applied to cloud resources. A user provisioning a cloud compute instance is generally billed for the time that the instance is used. You can add or remove the compute capacity based on your application's demands and only pay for what you used by the second, depending on the cloud provider you chose.

Measured service and reporting

The ability to measure cloud service usage is an important characteristic to ensure optimum usage and resource spending. This characteristic is key for both cloud providers and end users as they can measure and report on what services have been used and their purpose.

NIST states the following:

> *"Cloud systems automatically control and optimize resource use by leveraging a metering capability at some level of abstraction appropriate to the type of service (for example, storage, processing, bandwidth, and active user accounts). Resource usage can be monitored, controlled, and reported, providing transparency for both the provider and consumer of the utilized service."*

The cloud provider's billing component is mainly dependent on the capability to measure customers' usage and calculate the billing invoices accordingly. Cloud providers can understand the overall consumption and potentially improve their infrastructure's and service's processing speeds and bandwidth.

Businesses get the visibility and transparency they need to utilize their rates costs across large enterprises, which is limited in traditional IT environments. This is especially helpful for usage accounting, reporting, chargebacks, and also for monitoring purposes for their key IT stakeholders. In addition to the billing aspect, rapid elasticity and resource pooling feed into this characteristic, where end users can leverage monitoring and trigger automation to scale their resources.

In this section, we learned about the essential characteristics of cloud computing: on-demand self-service, elasticity, resource pooling, the PAYG model, measured services, CapEx/OpEx, and reporting abilities. In the next section, we look at what makes businesses inclined toward moving to the cloud.

Understanding the motivators for cloud adoption

The cloud has numerous offerings that help many organizations run their workloads on the cloud. By enabling cloud adoption, companies can accelerate their business transformations and expansions. Operating on the cloud helps companies classify and find motivations to help evaluate the necessities of migrating to the cloud. Let's look at some motivation-driven strategies that enterprises can expect as business outcomes upon performing cloud migration.

Resilience

The cloud's infrastructure is built on virtual servers that are built to handle substantial computing power and data volume changes. This helps cloud consumers leverage and build their applications so that they run without interruption. The cloud offers durable, redundant, pre-configured, and distributed resources that can be accessed from a variety of devices, such as laptops, smartphones, PCs, and more. The sophistication of this infrastructure allows you to build heterogeneous and multi-layer architectures that can withstand failures that are caused by unanticipated configuration changes or natural disasters when they're built right.

Having high levels of real-time monitoring and reporting capabilities in cloud environments to guarantee **service-level agreements** (**SLAs**) is nearly impossible for traditional data centers to build without substantial costs. This characteristic makes it easy for businesses to build robust and resilient applications with resource guarantees.

> **Service-level agreement (SLA)**
>
> An SLA is a measurement parameter (often expressed as a **percentage**) that defines a cloud service's expected performance and often serves as an agreement between the cloud service provider and the cloud consumer.

Note that cloud resiliency still requires businesses to build their critical systems with the right design, architecture, monitoring, orchestration, reporting, and governance to continue to run the businesses in the event of a disruption. However, with the cloud's underlying infrastructure, you can assess, evaluate, plan, implement, and manage your critical workloads and drive resiliency for your businesses as per your **recovery time objective (RTO)** and **recovery point objective (RPO)**.

> **What are RTO and RPO?**
>
> **RTO** is a business continuity metric that measures the amount of time a given application can stop working and the business can withstand the damage, as well as the time spent restoring the application and its data.
>
> **RPO** is a business continuity metric that measures the amount of lost data within a given period that is impacting the business, from the point of a critical event to the previous backup.

Advanced security

Cloud offerings, when used the right way with the proper security controls, can bring increased security to cloud consumers. The cloud service providers architect their infrastructures according to security standards and best practices to provide secure computing environments. The security-specific tools that are offered by the cloud service providers use controls to build their data center and network architectures, which are designed for high security and tightly restrict access to your data.

Carbon footprint reduction

Cloud computing continues to play a key role in reducing global energy consumption rates. Cloud computing is becoming an increasingly popular option for replacing on-premises server rooms and closets, which often lack the operational practices to consume energy efficiently, causing environmental impacts. The cloud enables organizations to share resources globally, resulting in higher efficiency and resource utilization compared to small private organizations that depend on standalone data centers.

As environment and climate awareness grows around the world, **cloud service providers (CSPs)** are continuously embracing and building their core physical infrastructure assets, which feed off of renewable energy. As the consumption of renewable energy increases, the overall carbon intensity will steadily decrease, resulting in energy transitions that help with the global climate and clean energy challenges.

At the macro level, cloud data centers invest in newer, more efficient equipment to achieve extremely high virtualization ratios, which are less likely to occur for typical enterprise data centers. The equipment's power consumption and cooling characteristics are an ever-evolving exercise that also helps reduce carbon emissions.

Improved optimization and efficiency

Cost savings is one of the key motivators for companies who are thinking of moving to the cloud. The setup and maintenance costs are usually reduced significantly by implementing cloud apps and their infrastructure. Surveys on cost savings and driving factors indicate that companies could save up to 50% on IT costs and cut down on the in-house equipment and the ongoing costs of maintaining IT departments with growing capacity needs.

Let's discuss a few factors that can drive cost savings:

- **Underlying hardware costs**: You don't need to invest in in-house equipment with cloud computing. This is a major cost cut for companies that don't have to worry about investing in upfront expenses to acquire underlying hardware and build on-premises server rooms or data centers. You can maximize the real-estate and office space, which also cuts down on costs.

- **IT operation costs**: You don't have to invest in employing any in-house staff to repair or replace equipment as this responsibility shifts from you to the cloud vendor when you migrate to the cloud. This is a major shift from capital expenditure to operational expenditure. You can free up your staff and focus on diversifying your workforce, who can work from anywhere with an internet connection.

- **Hardware maintenance**: Labor and maintenance costs are significant when it comes to building and maintaining an in-house data center. Ongoing upgrades or repairs are not your responsibility anymore, given that your data is stored offsite. This task will fall to the vendors, resulting in spending less time on installations from weeks or months to hours.

Moving to the cloud alone doesn't help with maximizing cost savings. You have to establish a cadence or a routine of monitoring cloud spending with the available tools by shutting down idle resources or rightsizing resources to realize extreme cost savings.

Faster innovation and business agility

With cloud computing, companies can become more business-focused than IT-focused and drive programs where the benefits matter the most. Some of these benefits are as follows:

- **Faster time to market**: Cloud-native offers end-to-end automation platforms that enable you to release code into production any number of times per day. As a result, businesses can adopt and bring new business use cases to the market about 40% faster.

- **Accelerates the innovation of business offerings**: Many popular cloud service providers have hundreds of native services in domains such as networking, databases, compute, **machine learning (ML)**, security, storage, **artificial intelligence (AI)**, business analytics, and many more. These can serve almost any industry, especially automotive, advertising and marketing, consumer packaged goods, education, energy, financial services, game tech, government, healthcare, and life sciences. Cloud offers a wide range of options for you to build, deploy, and host any application and this empowers companies to innovate rapidly.

In this section, we looked at why many businesses are moving to the cloud. We learned about the various factors that help them reduce IT operation costs and increase their business agility. Next, we'll learn about some of the leading cloud service providers and how their infrastructure is configured.

Understanding CSPs

When it comes to the on-demand availability and accessibility of cloud computing, CSPs offer these resources in many forms and sizes to businesses and individuals. Cloud consumers can rent access to any form of computing resources from applications to storage through these CSPs.

> **What is a CSP?**
>
> A CSP is a third party that offers on-demand cloud computing in the form of computing resources to other businesses or individuals without having them manage anything directly.

Some of the prominent cloud service providers across the worldwide cloud market are AWS, Microsoft Azure, Google Cloud, IBM Cloud, Alibaba Cloud, Salesforce, SAP, Rackspace Cloud, and VMware.

Let's take a look at a few of these cloud service providers and see what their offerings look like.

Amazon Web Services (AWS)

Launched in 2006, AWS is a cloud service provider that aims to offer a platform that is highly reliable and scalable. Over the years, AWS had strived to provide services that span geographical regions across the world. With over 170 fully-featured services, AWS is the world's most comprehensive and broadly adopted cloud platform.

Its service offerings feature across technical categories such as compute, databases, infrastructure management, data management, migration, networking, application development, security, AI, ML, and more.

As of 2022, AWS cloud spans 26 geographic regions and 84 availability zones around the world:

> **Regions and availability zones in AWS**
>
> An **AWS region** is a physical location around the world where data centers are clustered.
>
> Each group of logical data centers is called an **AWS availability zone**.

Figure 1: Magic Quadrant for Cloud Infrastructure and Platform Services

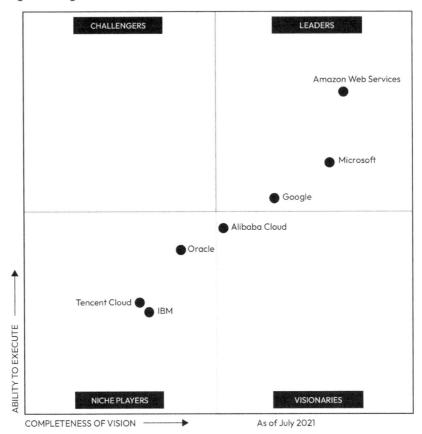

Figure 1.4 – Magic Quadrant for Cloud Infrastructure and Platform Services

The preceding diagram shows the magic quadrant for the cloud infrastructure that was published by Gartner in 2021.

> **Note**
>
> Each CSP has terminology to indicate the cloud regions for the consumer's needs based on technical and regulatory considerations.

Microsoft Azure

Launched in 2010, Microsoft Azure is one of the fastest-growing clouds and offers hundreds of services across categories such as AI, ML, analytics, blockchain, compute, databases, and more.

Azure's global infrastructure is made up of two key components – **physical infrastructure** and **connective network components**. The physical component comprises 200+ physical data centers, arranged into regions, and linked by one of the largest interconnected networks on the planet (source: `https://docs.microsoft.com/en-us/azure/availability-zones/az-overview`).

As of 2022, Azure consists of over 60 regions worldwide across 140 countries.

> **Regions and availability zones in Microsoft**
>
> A **region** is a set of data centers that are deployed within a latency-defined.
>
> Unique physical locations within a region are called **availability zones**. Each zone is made up of one or more data centers.

Google Cloud Platform

Launched in 2008, **Google Cloud Platform** (**GCP**) is a suite of over 100 products and services offered by Google. Its core service offerings include compute, networking, storage and databases, AI, big data, identity and security, and more.

As of 2022, Google Cloud spans over 28 cloud regions, 85 zones, and 146 network edge locations across 200+ countries and territories.

> **Regions and zones in Google**
>
> Each data center that has a location that comprises physical assets such as virtual machines, hard disk drives, and more is defined as a **region**.
>
> Each region is a collection of **zones** that are isolated from each other within the region.

Alibaba Cloud

Founded in 2009, Alibaba Cloud's wide range of high-performance cloud products include large-scale computing, networking, databases, storage security, **Internet of Things** (**IoT**), media services, and more.

As of 2022, Alibaba Cloud operates around the world with over 78 availability zones in 24 regions.

> **Regions and zones in Alibaba**
>
> A **region** is a geographic area where a data center resides.
>
> A **zone** is a physical area with independent power grids and networks in a region. The network latency for access between instances within the same zone is shorter.

In this section, we looked at some of the popular companies that are managing cloud computing through their cloud technology offerings. Next, we will provide an overview of the cloud service models and discuss the level of management each model provides.

Exploring the service models – SaaS, PaaS, and IaaS

As you navigate your path to the cloud, there are key decisions that you must make that revolve around how much you want to manage yourself and how much you want your service provider to manage. These cloud service models can be put into three categories that match your current needs so that you're prepared for the future:

- **Infrastructure as a Service (IaaS)**: IaaS is a service model that offers consumers on-demand access to virtualized compute, storage, and networking.

- **Platform as a Service (PaaS)**: PaaS is a service model that offers consumers on-demand access to a ready-to-use cloud-native platform for developing, running, hosting, managing, and maintaining applications.

- **Software as a Service (SaaS)**: SaaS is a service model that offers consumers on-demand access to ready-to-use software for cloud-hosted applications.

The following diagram shows what you manage for each type of model:

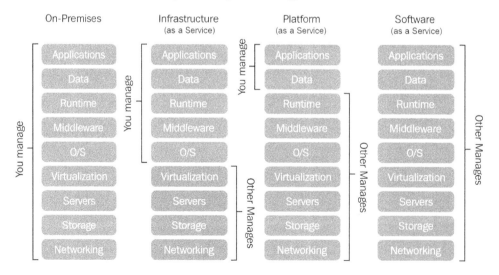

Figure 1.5 – Cloud models

Let's discuss each of these in more detail.

Infrastructure as a Service (IaaS)

Infrastructure services enable companies to acquire resources on-demand and as-needed. This gives users cloud-based alternatives instead of them having to buy the required hardware, which is often expensive and labor-intensive. The main offerings include computing resources such as storage, servers, and networking.

The features of IaaS are as follows:

- Highly flexible and highly scalable
- On-demand offerings that can be accessed via the internet
- Highly redundant as data lives in the cloud
- Zero management is needed for the virtualization tasks
- Cost-effective

Ease of use

As shown in the following diagram, IaaS is often called an **everything-as-a-service** business model:

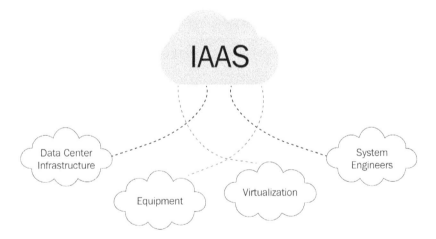

Figure 1.6 – The IaaS model

IaaS is suitable for companies of any size and complexity. However, there are some use cases where companies can find IaaS more beneficial:

- **High-performance computing**: Performing groundbreaking complex calculations for batch processing workloads, media transcoding, scientific modeling, and gaming requires a high-performance computing architecture with clustered compute servers and data storage. IaaS can be leveraged to take advantage of its rapid scalability and support for networking compute resources.

- **Disaster recovery and backup solutions**: Building a disaster recovery plan on-premises involves a complex infrastructure that requires fixed capital expenses. With IaaS, this can be achieved in a few steps; all you need to do is set up the required infrastructure services for disaster recovery and backup solutions.

- **Real-time data analytics**: The ever-increasing requirement of applications to analyze data in real time requires decisions to be made in seconds. Collecting real-time data and processing it can be a time-consuming and expensive development endeavor.

IaaS can be used to manage, store, and analyze big data and handle large workloads while easily incorporating business intelligence tools. Getting business insights out of this raw data and predicting trends can be rendered effectively.

You should consider the following factors if you wish to choose an IaaS provider:

- **Security**: Protecting sensitive data, standardizing identity management procedures, and evaluating compliance standards are some of the security procedures that can dramatically impact your security posture when you're using an IaaS model. It's important to make sure that the IaaS provider is protected against security risks.

- **Pricing model**: In addition to the initial expenses, make sure that you understand your IaaS provider's pricing structure and the different monitoring tools and mechanisms that they are providing for monitoring and tracking your spending. Sometimes, the initial pricing may convince you to migrate from your on-premises infrastructure, but laying out a long-term plan with expected savings will enable you to plan for resource provisioning effectively.

- **SLA and support process**: Knowing your vendor's SLAs to ensure any infrastructure issues are resolved promptly is crucial for your businesses to run without interruptions. Understanding the level of support they provide once you become a paying customer is crucial.

- **Integration capabilities**: When you're migrating to an IaaS model, it is important to understand how your current workflows can be incorporated into the cloud without major customizations. Without proper integration, your products may suffer from additional development and administrative efforts, which will often translate into higher costs and application support.

- **Latency requirements**: Analyzing which IaaS will provide you with the closest/less latency to your customers is also important. Also, if the data you need must be located physically in a country, make sure that the IaaS provider has facilities in that country. The following questions can help you address this:

  - Where are this IaaS provider's closest data centers?

  - How many data centers are in the region I'm interested in?

  - Is there a region/data center/facility in the country that my data needs to live in?

In summary, IaaS represents general purpose compute resources to support customers-facing websites, web applications or customers that are heavy on data, analytics and warehousing. IaaS supports a diverse set of workloads and, as we explore in later chapters, we will look into the emerging compute models that are positioned for modern application architectures such as microservices.

Platform as a Service (PaaS)

Platform services enable developers to build applications by providing hardware and software tools that can be accessed from the internet. Businesses have the freedom to incorporate special software components while they are designing and creating applications. The cloud's inherent characteristics enable these components to be highly scalable and available. PaaS provides application life cycle management tools and **integrated development environments** (**IDEs**) to help you select the best for your needs.

An important characteristic of PaaS is that it lets you manage how different tenants are isolated. So, if the load on one tenant becomes high, the demand is distributed to the right instances of the applications. This function enables high scaling and availability. Developers can build applications anywhere in the world and don't have to worry about operating systems, storage, the underlying infrastructure, or software updates:

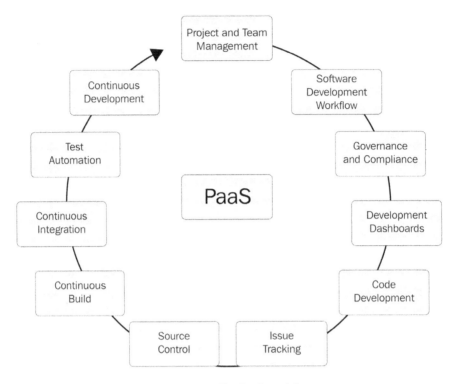

Figure 1.7 – The PaaS model

The features of PaaS are as follows:

- Developer-friendly
- Built on virtualization technology
- Scalable and highly available
- Quicker churn in coding and testing
- Pluggable customizations
- Available to multiple developers at the same time
- Easy integration of web services and databases
- Lower capital commitment
- You can manage applications easily
- Reusability

With PaaS, you get support for application development, operating software, deploying to IaaS infrastructure automatically, handling runbook scenarios automatically, and bringing many improvements, including monitoring your end-to-end applications. Let's look at why companies usually implement PaaS:

Many forward-thinking companies, including large businesses, small start-ups, and everything in-between want to build projects and create an open source-like world where everybody inside the company can have access to the code of all the other projects to reuse. They are hoping that they will be able to use common code and services to increase their productivity and innovation.

With PaaS, developer efficiency can be tremendously enhanced by leveraging common tools so that they can realize the benefits of the open source technology approach.

A key success criterion is moving as many teams and projects to the new PaaS as quickly as possible. This ensures that there's sufficient mass within the company to create the desired innovation and shared development benefits such as the following:

- **Cut down on costs**: Many companies invest large amounts of capital to run existing legacy applications where they would like to reduce the ongoing spending. To run mainframe-like applications, companies need to invest in special hardware or build the virtualization themselves. Legacy applications cannot be changed because of resource constraints such as the original owner being gone or out of the business. With PaaS offerings, you can reduce specialized management, increase the ability to share resources, and cut the costs of operating these applications by 50-80%.

- **Increase reusability**: Many companies are looking for ways to build several APIs or applications as quickly as possible. PaaS provides the required middleware so that you have a reusability paradigm where developers can design and build common features that can be reused.

- **Faster time to market**: Due to its cost-effectiveness and access to state-of-the-art resources, companies of any size can build robust systems at a faster pace and accelerate their launch times.

The idea behind reusability is to use something that already exists rather than investing money and resources into creating something new. This is especially critical for new companies that are starting from scratch and it is even more important for their businesses to do things right. Reusability is an important aspect, where the developers can reuse existing APIs, existing services, or components that will help you to start with low costs and develop faster. There are a few factors to consider when you're choosing the right PaaS provider for you:

- **Developer support**: Learning whether your PaaS vendor providers platforms that support all major programming languages, easy-to-use templates to build and deploy applications, and support for relational and non-relational databases is recommended. Researching how well-equipped your vendor is when it comes to streamlining the developer's processes and deployment procedures is also helpful before locking in on a provider.

- **Compliance and Regulation**: Vendors that adhere to the industry standards and regulatory requirements will be helpful for you to be in alignment with the best practices in the industry.

- **Reliability and Performance**: Your potential PaaS provider should be implementing disaster recovery and fault-tolerant techniques to ensure the RTO and RPO of your business systems. An ideal provider will have strategies and processes to guarantee both planned and unplanned events.

- **Data Security**: Data is the heart of any application and providers need to be able to guarantee SLAs, and support guidelines that will respect the confidentiality of ensuring data security and privacy.

In summary, the PaaS market is competitive and offers unique solutions for customers of any size or vertical to build customized applications faster.

Software as a service (SaaS)

SaaS is the most common category of cloud computing that enables users to leverage software through the web or APIs. SaaS offloads the users from downloading or installing applications on their local devices that they like to use in collaboration with their projects. Storing and analyzing the data is also done via a remote cloud network. The SaaS solutions are hosted centrally in the cloud and can be accessed from anywhere, anytime.

The vendor is fully responsible for the consumer's software experience. Users are not responsible for hardware management, nor the functionality of these vendor-hosted applications. Many popular solutions such as Salesforce (customer relationship management software), Canva (graphics), and Slack (collaboration and messaging) are all examples of SaaS.

The following are some of the features of the SaaS model:

- **Reduced risk**: Companies don't have to build the software that meets their needs; they can try out the SaaS products for a zero or low fee. This characteristic enables them to explore and evaluate the best product that can be used with their existing system with no financial risk.

- **Increased productivity**: SaaS products, as opposed to the traditional model, can enable users to provision their resources and start using them in a few hours. This cuts down on the installation, configuration, and deployment time so that the users can focus more on building their applications and innovate faster.

- **Increased data protection**: Data is routinely stored in the cloud and all software maintenance in the form of updates, version upgrades, and support is taken care of by the service provider. As a result, data loss is protected from hardware failures.

- **Customers don't manage, install, or upgrade the software**: The provider takes care of making upgrades available to their customers.

- **Easy to access**: Applications are accessed via internet-connected devices. Analytic tools for data reporting and intelligence tools, which are often expensive and difficult to configure, are easily available.

- **Flexible pay model**: Customers pay to gain access to the software and applications. Start-ups and small businesses can easily leverage SaaS applications, given that they are easy to set up and no capital or expertise is required to build these tools and applications. There is a variety of business scenarios where you can take advantage of the SaaS model and its self-provisioned nature.

- **Applications with web and mobile access**: Companies that are building applications that require web and mobile access can take advantage of the SaaS technology.

- **Short-term projects**: Companies often use SaaS technology when they're building applications that are short-term projects and that are not required all year long.

- **Startups and small businesses**: You will find that SaaS comes in handy when there is not much time, capital, or expertise available to build your applications and host them on-premises.

While choosing the right SaaS vendor, make sure that you consider the following factors:

- **Limited customizations**: SaaS applications are available out of the box with limited customizations from the vendor. If you have a strong requirement that is dependent on the SaaS application, you may want to consider building some customizations to work around this limitation.

- **Limited compliance**: Vendors own the SaaS applications and continually release new features and fixes. However, you lose a degree of control when it comes to meeting ever-increasing legal, regulatory, and compliance requirements to keep your organization away from non-compliance. Make sure that you ask your SaaS provider about uptime, resiliency, and any critical compliance requirements ahead of signing up.

- **Vendor viability**: Acquisitions are commonplace for the SaaS market and often result in terminations of services with very short notice. Ensure that you agree on an SLA with an uptime guarantee and also have a thorough understanding of the duration of your contract.

In this section, we looked at various cloud computing models and their use cases, benefits, and limitations. Next, we will discuss the cloud deployment models in detail

Exploring the deployment models – public, private, hybrid, multi, and community

As the cloud is increasingly becoming the default option for many companies, you must choose the cloud model that is most suitable for your needs. Choosing a cloud environment type and a deployment model that aligns with your business goals is a process that you can dive into before you start your cloud journey.

> **Cloud deployment model**
>
> The cloud deployment model is defined by a combination of deployment types that control parameters such as the accessibility, location, and proprietorship of the infrastructure, network, and storage size.

When it comes to cloud deployment models, there are five main types:

- Public

- Private

- Hybrid

- Multi

- Community

We will discuss each of them in detail in the following subsections.

The public cloud

The public cloud deployment is accessible by anyone and is the most commonly used model. The main feature of this deployment type is that you don't know or own any hardware. The service providers manage the server infrastructure for you, they administer the resources and maintain the hardware, and you are charged on a pay-per-use basis in most cases. Data is created and stored on the servers and these servers are shared between all the consumers:

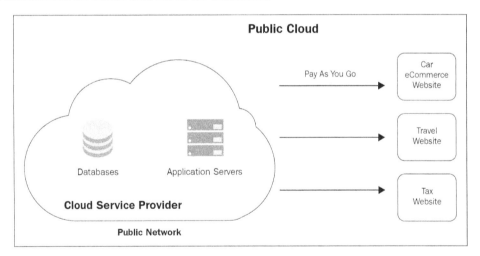

Figure 1.8 – The public cloud

There are many benefits that businesses can reap from a public cloud deployment model:

- **Easy to set up**: Most CSPs have intuitive and easy-to-use portals to set up resources.

- **No infrastructure maintenance**: The CSP is responsible for maintaining the underlying infrastructure.

- **Elasticity**: It's easy to acquire or release resources to meet your business requirements.

- **Highly available**: The extensive ecosystem of your provider's resources provides the required controls to run your workloads with improved uptime.

- **Cost-effective**: There's a PAYG model for the services that you use and no upfront investments to purchase hardware or software.

Apart from these benefits, there are a few factors to consider if your requirements resonate with the following points:

- **Security and risk mitigation**: While the CSPs implement many mechanisms to make the cloud highly secure, your applications and data in the public cloud are only secure with your help. Many CSPs come with native encryption, automation, access control, orchestration, and endpoint security mechanisms to manage risk effectively.

- **Prone to large-scale infrastructure events**: Cloud service providers strive for high availability, but public clouds have suffered outages in the past that caused huge losses. You must do your research before deciding on the cloud computing provider for your applications that need to have the highest uptime. Irrespective of the cloud provider you choose, it is important to have an enterprise-wide incident management and remediation platform strategy to handle events effectively.

- **Standard features a "one-size-fits-all" model**: Many cloud service providers offer a standard set of features that cater to most companies. However, you will need to consider additional customizations or workarounds if you have applications that require CSPs to develop complex features. Some primary examples of public cloud models include **Amazon Web Services** (**AWS**), Microsoft Azure, IBM Cloud, and Google Cloud.

The private cloud

The private cloud, as its name suggests, is a dedicated cloud model where a specific business or company owns the private cloud. While the architecture of the public and private cloud is similar, the difference is in the way you own and manage the hardware. Most commonly, the hardware will be dedicated to you and you don't share it with any other users outside your company.

The service provider will provide you with an abstraction layer for all the hardware. Here, you will be able to add new hardware to your cloud but will not be responsible for configuring it, given the semi-automatic nature of the provisioning process. You may choose this model when you have stringent security and compliance restrictions regarding the nature of the applications that you may want to run and are ready to pay high costs for the dedicated setup:

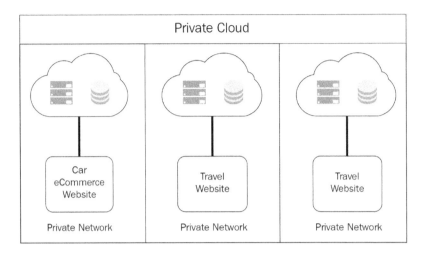

Figure 1.9 – The private cloud

Some of the characteristics of the private cloud model are as follows:

- **Increased security**: Cloud access through private and secure network links, along with the native antivirus, firewall protection, and encryption mechanisms, makes the private cloud environment more secure.

- **Increased regulatory compliance**: Due to its security and control benefits, the private cloud can help address regulatory compliance hosting requirements.

- **More flexible infrastructure model**: Many organizations that are moving their workloads from legacy on-premises to the cloud find it difficult to meet the customization requirements that support their applications. The infrastructure of the private cloud can be configured to provide services and support for such stringent requirements.

While there are many benefits of using the private cloud model, make sure you learn more about the following limitations before choosing the private model:

- **Increased costs**: The private cloud model can be more expensive than the public cloud because of the infrastructure expenses that you have to spend.

- **Maintenance and deployment**: Continuous deployment and maintenance require additional setup and staff, which can be time-consuming.

- **Limited remote access**: The private cloud has limited remote access, so mobile users may not be able to connect to the cloud whenever they want. Some examples of CSPs that provide the private cloud include Amazon, IBM, Cisco, Dell, Red Hat, Rackspace, Microsoft Azure, Red Hat OpenStack, and VMware.

The hybrid cloud

The hybrid model is a combination of on-premises, private cloud, and/or public cloud services that lets you get value from all the features of all the models. This model allows you to mix and match the other models' capabilities to best suit your business requirements.

The hybrid cloud deployment model facilitates data and application portability to safeguard and control your assets strategically. Being able to balance multiple deployment models not only safeguards the controls but helps maximize the benefits of cost and resource utilization. Many organizations are evaluating this as a transitional model that eases you into the public cloud over a longer period:

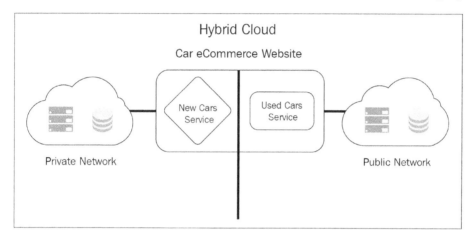

Figure 1.10 – The hybrid cloud

The following are the benefits of the hybrid cloud model:

- **Improved speed**: The mobility between cloud models gives you greater speed and agility for innovation and speed to market. You don't have to be limited to your private on-premises environment and can expand your workload quickly to test, prototype, and launch new solutions.

- **Business continuity**: The hybrid model helps reduce potential downtime and impacts in the event of a failure or a disaster. You get improved business continuity and can continue with business operations when you opt for the hybrid model as a backup option during interruptions.

- **Improved security and privacy**: Due to security restrictions or data protection requirements, few companies cannot operate only in the public cloud. This model provides an improved security model platform for mission-critical applications with sensitive data on-premises while you're running the remaining applications in the public cloud.

- **Improved risk management**: You get more control over your data and improved security, which means you can reduce data exposure. You get to standardize cloud storage and implement stronger security controls to manage risk effectively.

The following are a few limitations to consider before choosing the hybrid model:

- **Managing multiple vendors and platforms**: You will have to keep track of and manage multiple vendors and platforms to have effective computing environments. Having runbooks, workflows, and processes with a good team understanding and effective coordination of vendors is a must to make sure your environments are running without interruptions.

- **Hardware costs**: The cost that's associated with hardware procurement, setup, maintenance, and installation of the hybrid cloud infrastructure is high. Organizations will have to prepare for this upfront cost, as well as train their IT staff to cope with the cloud and on-premises expenses.

- **Security**: On-premises and the cloud require different approaches to secure your applications. Using a blend of public, private, and/or on-premises makes it difficult to be free of intrusion risks.

- **Lack of visibility**: Hybrid increases the number of environments that the operations teams need to keep track of and achieve a clear view of. Management becomes difficult if you don't have a good understanding of the current infrastructure and operations, which leads to missed opportunities regarding potential issues.

Multi-cloud

Cloud providers recommend many design patterns to achieve high availability for applications running on the cloud. When you are using more than one cloud provider at a time to achieve high availability, you are using the multi-cloud deployment model. Companies may also use the multi-cloud option when they need a specific service from a CSP X and another specific service from a CSP Y.

The multi-cloud approach involves adopting a mixture of services from multiple cloud providers, sharing workloads between them, and picking services that meet specific business needs to achieve greater flexibility and reliability:

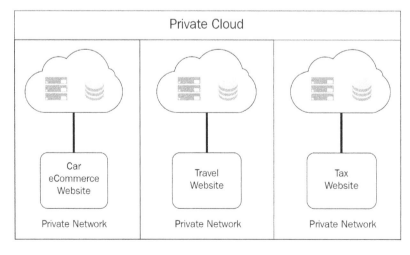

Figure 1.11 – Multi-cloud

The following are the advantages that businesses can reap while using the multi-cloud approach:

- **Multiple best-in-class cloud providers**: Each cloud provider has its strengths and weaknesses when it comes to providing features that you need to use for your applications. The foremost benefit of the multi-cloud strategy is the ability to take advantage of the unique best-in-class services that each cloud provider offers. You get to pick and enable your developers to focus on innovation and unblock any limitations that a specific cloud provider may have.

- **Avoid vendor lock-ins**: Many businesses worry about getting locked into a specific cloud provider or infrastructure and pricing model when using a single cloud provider strategy. You have greater flexibility in choosing the multi-cloud to leverage the best of the services that the cloud providers offer. You get to pick the vendor that has a specialized and evolved set of services.

- **Risk mitigation and enhanced resiliency**: Continuous availability is a key aspect for any business that runs mission-critical workloads. You get the option to run your applications and store data on multiple clouds to fall back on and restore in the event of a service outage.

- **Flexibility and scalability**: Multiple cloud vendors invest in a higher amount of space, security, and protection to offer a perfect place for your businesses to process and store information. With the right expertise at hand and having a good multi-cloud operations runbook, you can achieve greater scalability, which allows your applications to scale the storage or compute up or down based on the ongoing demand.

Although there are many advantages of using multiple cloud vendors, building and managing a multi-cloud architecture can have its downsides:

- **Building the expertise**: The need for cloud computing expertise is growing at a rapid pace. Many companies are having trouble recruiting cloud professionals that have the knowledge and skillset of a single cloud provider. It is a challenge to find network specialists, security experts, architects, and engineers that have expertise in multiple clouds. Within your organization, you will need to plan out how you will recruit the right workforce and develop their skill set on multiple cloud platforms to build, secure, manage, and operate your applications across multiple clouds.

- **Cost tracking and optimization**: Each cloud provider has a specialized set of tools and reporting platforms to help you manage the financial costs of your resources running on their cloud. Consolidating these costs and having a good handle on their pricing model to navigate through the math and pricing structures is recommended when you're operating on multiple clouds.

- **Increased complexity on operations**: Many companies find moving to the cloud a long and daunting task. In addition to that, managing workloads on multiple clouds may add to the complexity if it's not planned well. With your applications and their resources spread across multiple clouds, operational management such as patching, monitoring, logging, and backing up your resources are all details that you have to consider when planning.

- **Security risks**: It is important to understand the blast radius of security attacks when it comes to applications that are deployed on multiple clouds. Considering how well you configure, manage, alert, log, and respond to such security breaches must be accounted for. Many companies use third-party tools to manage their approaches on encryption keys, identity and access controls, and resource policies.

- **Compliance**: Creating a shared responsibility model with multiple cloud providers can be a daunting task. Simplifying how vulnerabilities are managed and solving compliance challenges can be a few drivers that can add complexity.

The community cloud

The community cloud, although less popular than the previously discussed models, is a hybrid form of a private cloud that has a similar architecture and the ability to use security and privacy controls. Organizations get to run their workloads on a shared platform where multiple consumers can work on projects and applications that may belong to specific industry segments. Businesses such as health care companies, financial institutions, governments, research, education, and even large manufacturing companies are ideal industries for community cloud environments:

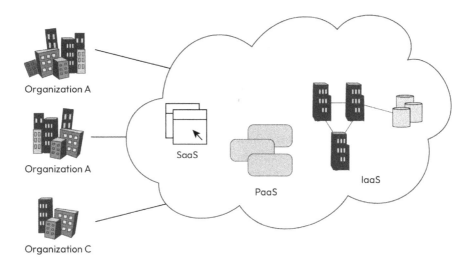

Figure 1.12 – Community cloud

Let's look at some of the advantages of adopting a community cloud strategy:

- **Convenience and control**: The community cloud offers the same flexibility as a public cloud environment and has the same security levels and privacy as a private cloud. This makes it accessible for a specific set of organizations and gives you much more confidence in the platform, as you can govern your applications with industry-tailored flexibility.

- **Security and privacy**: The community deployments are similar to that of the private cloud, where you can control security at more granular levels. This ensures that secure transactions align with regulatory protocols.

- **Availability and reliability**: Community clouds provide the same level of services to ensure the availability of your data and applications at all times. Replicating your data and applications in multiple locations enables you to implement redundant infrastructure for your critical applications where availability and reliability are topmost priorities.

However, there are a few concerns regarding the community cloud approach that you will want to identify and evaluate before adopting this model:

- **Limited storage and bandwidth**: Data storage and bandwidth are shared among other organizations, which limits the community members to a finite amount of data storage and bandwidth.

- **Not a "one-size-fits-all" model**: The community cloud approach is a new model that has recently started evolving as more and more businesses are finding a fit for their use cases. Small, medium, and large businesses must still evaluate this on a case-by-case basis, given that many public cloud providers are offering services that cater to the requirements of every business.

Comparison between the different cloud deployment models

The following comparison matrix shows how each model fairs when it comes to parameters such as security, risk management, reliability, scalability, and cost:

Parameters	Public	Private	Community	Hybrid	Multi
Security	Depends	High	High	Medium	Depends
Privacy	Depends	High	High	Medium	Depends
Risk Management	Depends	High	High	Low	Depends
Reliability	Depends	Medium	Medium	Depends	Depends
Scalability and Flexibility	Depends	High	High	High	Depends
Cost	Low	High	Depends	High	High
Data Control	Medium	High	Low	Depends	Depends

Table 1.1 – Deployment model comparison matrix

The preceding table can be used as a cheat sheet as you evaluate various deployment models and determine which model will be best suited for your business's requirements.

Summary

In this chapter, we introduced the cloud and some of its concepts, such as on-demand self-service, resource pooling, multi-tenancy, elasticity, and scalability. We learned about the history of the cloud and discussed the key motivators for businesses to move to the cloud. We categorized the cloud service models and deployment models before looking at some of the commonly used cloud vendors and learning about their infrastructure in detail.

After that, we discussed when we should use a specific cloud model and the factors to consider while choosing a specific model. We familiarized ourselves with concepts such as public, private, hybrid, multi, and community cloud models while looking at each model's benefits and additional factors to consider.

These concepts should have given you an in-depth understanding for the next chapter, where we will focus on cloud migration fundamentals and the different phases of cloud migration.

Further reading

For more information on cloud fundamentals and various cloud providers, please read the following articles:

- Amazon Web Services – `https://aws.amazon.com/about-aws/`
- Global Infrastructure of AWS – `https://aws.amazon.com/about-aws/global-infrastructure/`
- Microsoft Azure – `https://azure.microsoft.com/en-us/`
- Global Infrastructure of Azure – `https://azure.microsoft.com/en-us/global-infrastructure/`
- Availability Zones with Microsoft – `https://docs.microsoft.com/en-us/azure/availability-zones/az-overview`
- Google Cloud – `https://cloud.google.com/`
- Google Cloud Overview – `https://cloud.google.com/docs/overview`

Understanding Cloud Migration

With the previous chapter, we were able to establish how vital the cloud is to achieving a transformational shift and enabling you to become *cost-efficient, agile, and innovative* in how you operate your businesses. However, many enterprises experience *fear, uncertainty, and doubt* when they are at the beginning of their cloud migration journey, resulting in resisting the change. Some industry segments see the cloud as a hindrance in that they are concerned about the cost and effort it takes to move to the cloud, while others are hesitant about storing data from a compliance or security perspective.

This chapter will give you an understanding of **cloud migration** and what is involved to make the transition happen and will provide a detailed navigation to overcome the roadblocks that can potentially slow down your transition to the cloud. Using this as a guide, you can build a mental model to take on cloud migration efforts with an effective understanding of its fundamentals.

In the context of this book, I will be talking about AWS offerings wherever applicable to enable you to have a better understanding of how to be successful with your migration and modernization efforts on AWS.

In this chapter, we'll cover the following topics:

- The key concepts and fundamentals of cloud migration
- Understanding the key migration challenges
- The evolving benefits of cloud migration
- Building a cloud-first mindset
- Exploring the phases of cloud migration

Cloud migration – key concepts and fundamentals

Understanding cloud migration is crucial for businesses of all sizes to make their cloud journeys effective. Before we take a deep dive into the activities needed for cloud migration, let's understand what cloud migration is.

What is cloud migration?

Let's do a quick recap on cloud migration from the previous chapter.

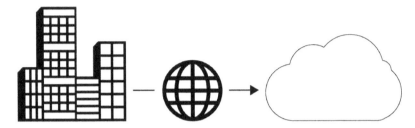

Figure 2.1 – Cloud migration

Cloud migration is the process of taking your digital footprint and assets – such as software applications, data, IT resources and infrastructure, and workloads running either on-premises or on the cloud – and moving them to a (different) cloud.

Workloads can be either applications, databases, storage, physical or virtual servers, or websites.

A migration typically means moving a few or all assets to a cloud and involves many things.

Typically, many businesses run hardware or software that is in use but outdated and is considered **legacy**. Legacy applications are often at risk of being unreliable, with performance bottlenecks, or using software that is no longer supported by its vendor. Legacy hardware typically includes aging equipment or servers that support the company's business-critical software, which adds security risks and availability issues. Enterprises that rely on such legacy infrastructure are propelled to consider cloud migration in order to benefit from the cloud's inherent scalability, performance, flexibility, and cost savings.

Migrating to the cloud safely typically involves meticulous planning and deliberate execution. Although most of the time, cloud migration describes the transition from legacy on-premises to the cloud, this may also apply to moving from one cloud to another cloud.

AWS cloud fundamentals

Before we dive deeper into the cloud migration challenges and their phases, let us get started with some of the core concepts of a cloud-native paradigm. For many, the cloud can be a radical change, so understanding the basic concepts of the cloud will help you to apply them across any service that AWS offers.

I will cover the six main pillars of AWS – Operational Excellence, Security, Reliability, Performance Efficiency, Cost Optimization, and Sustainability. As you read through these concepts, try to build a mental model using these concepts – that will become your foundational knowledge as you get familiar with AWS's services and technologies.

Operational Excellence

Operational excellence is the ability to improve how you run your systems by creating automation. Two areas are directly related to operational excellence:

- **Infrastructure as Code** (**IaC**) – Provisioning and managing your infrastructure through declarative code files is the key concept behind IaC. Some of the services available in AWS are AWS CloudFormation and AWS **Cloud Development Kit** (**CDK**).

- **Observability** – Observability is measuring the state of your systems to optimize and improve them continuously. Collecting, analyzing, and actioning the metrics that are captured at either the service or application level is the key driving process of observability. AWS offers full-stack observability, including **application performance monitoring** (**APM**), and open source solutions for understanding your technology stack at any time. Amazon CloudWatch lets you observe your resources on AWS and on-premises. You can do distributed tracking across multiple applications using AWS X-Ray. Amazon Managed Services for Grafana and Prometheus lets you do data visualization and monitoring for your workloads.

Security

Security is the pillar that provides tools and mechanisms on how to best secure your workloads on the cloud. Security is a shared responsibility between AWS and the customer. As we discussed in the first chapter, AWS is responsible for the security of physical facilities, and for its infrastructure, software, and networking capabilities. The customer is responsible for their application software and their data.

A **zero trust model** is encouraged by AWS where security measures at all levels need to be applied to all application components and services. **Identity and access management** (**IAM**), network security, and data encryption are three broader categories in securing systems with zero trust in AWS.

AWS services such as **IAM** and AWS **Web Application Firewall** (**WAF**), including network level primitives such as Amazon Virtual Private Cloud, security groups, and proxy services such as application load balancers, are offered to apply security measures as required.

Reliability

The reliability pillar enables you to build applications that are equipped and resilient for both service and infrastructure disruptions. The **blast radius** is a key concept when it comes to thinking about the maximum impact that a system failure can have on your components, causing them to be down or unavailable.

Fault isolation and service limits are two techniques to limit the blast radius when a failure happens. AWS provides fault isolation zones at the region/availability zone/resource level, where it uses techniques such as **shuffle sharding**.

> **Shuffle sharding**
>
> Shuffle sharding is the process of creating virtual shards, assigned to customers or resources; upon a service impact, it's easier to keep the impact smaller and handle most of the scaling challenges easier. This powerful yet simple core concept is how AWS is able to deliver a cost-effective and multi-tenant experience to its consumers.

Amazon Route53, Regions, Availability Zones and Local Zones, Amazon Service Quotas, and AWS Trusted Advisor are some of the service offerings that are categorized in this pillar.

Performance efficiency

Focusing on how efficiently you can scale your applications on the cloud, performance efficiency enables you to think of applications and services as **cattle not pets**.

Pets are often treated as unique and requiring a lot of maintenance, whereas the **cattle model** is considered cheap and fast, which gives businesses the benefit of the freedom to select the services that best match their requirements.

Cost optimization

The pay-as-you-go model is a fundamental concept of the cloud and changes your approach to how you want to build and deploy your workloads cost-effectively on the cloud. Cost optimization is a continuous iterative cycle where you can achieve your business outcomes while reducing costs as you operate.

Some of the cloud concepts for cost optimization are right-sizing, serverless, reserved compute, and spot compute to pay as effectively as you can. Reviewing, tracking, and optimizing your cloud expenditure are the three key workflows that you can use to achieve tremendous cost optimization results.

AWS Cost Explorer, AWS Cost and Usage reports, and AWS Budgets are some of the offerings for this pillar with which you can organize and report cost and usage effectively. Cloud financial management is a very important aspect, and we will discuss this in detail in the coming chapters, as well as how to excel in cloud cost strategies while running on AWS.

Before now, we have gone over the fundamentals of cloud migration and discussed how aspects such as operational excellence, security, and cost play a key role. In the next sections, we will look into factors that can be blockers for cloud migration and how to tackle them effectively.

Sustainability

Reducing the environmental impacts of running the workloads on the cloud is the main focus of the Sustainability pillar. Through the shared responsibility model for sustainability, you can understand the impact and importance of maximizing your utilization and reducing downstream impacts.

This pillar addresses aspects of the environmental, economic, and societal impact of your company's business activities. Environmental impacts such as carbon emissions, unrecyclable waste, and damage to resources such as clean water are increasing; it is critical to reduce resource usage to address such impacts. Energy consumption and efficiency become important levers while applying design principles and best practices to result in the reduction of such impacts.

Understanding the key cloud migration challenges

Many organizations face challenges during cloud migration, which can often lead to concerns such as delays in time to market, cost overruns, and security vulnerabilities. These challenges result in service outages, data losses, and connectivity issues.

The reasons for these challenges often stem from not having good control over aspects such as the following.

Lack of enterprise-wide vision

Businesses that do not have good visibility into applications, users, and network traffic are one of the demographics that struggles the most. This visibility is an important aspect of cloud migration as it helps businesses identify the critical components of their workloads that need to be migrated first to ensure that there is no downtime.

Many CTOs have a difficult time getting a mapping for each of their applications and dataset and the business metrics associated with them to prioritize a migration accordingly.

Migrating production-grade data or applications without knowing how their workloads will run in a cloud environment can add to the migration bottlenecks.

Answer the following question before proceeding any further with your cloud migration: *What is your cloud vision?*

Lack of cloud-first mindset

Many seasoned and successful technology executives are stuck with the old wisdom of their legacy applications, which can be an inhibitor to their future successes. Many companies struggle with adopting a cloud-native mindset and are challenged to rethink their end-to-end digital transformation.

Putting the cloud in context with the future of your workloads is a paradigm shift and is a forward-looking enterprise strategy, impacting the way you will run your businesses. To meet the next-generation demands, it is important to have the vision to extract the value of your business's investments through the cloud and enable your teams to develop their applications so that they can hit the ground running with the latest and greatest.

Lack of technical skill sets

The skills that are required to develop and manage on-premise servers and applications are different from those involved in running workloads on the cloud. A cloud talent shortage is a leading business concern that slows down cloud migration projects.

Many businesses start their cloud migration without recruiting the required talent or ensuring their existing employees are trained to work with cloud technologies. A lack of qualified and skilled professionals to build and manage the workloads can result in a massive cloud migration challenge.

The 2021 Forrester study suggests that cloud ecosystem expertise is one of the top skill shortages that can take longer to master, and companies need to invest in continuous training.

Make sure your organization can allocate the funding for the following:

- Training
- Staffing additional technical resources

Lack of a well-defined strategy

Many businesses find it overwhelming to create detailed guidance or instructions on how to align various teams across their organization that informs them about cloud service adoption. Determining the key benefits of cloud service consumption relevant to your business and identifying the services and partnerships form the crux of the strategy.

It is crucial that leaders and architects of businesses build a well-crafted strategy that talks about the *why*, *how*, and *what* of cloud migration to guard the execution with the least risk exposure.

The absence of a cloud strategy can lead to an increased risk of unproductive cloud migration and greater risks of siloed execution, resulting in poorly configured environments. Developing a high-level plan and defining metrics for the plan are encouraged for a seamless migration. Organizations can focus their planning on the following for successful cloud migration:

- An enterprise-wide comprehensive cloud migration strategy
- An understanding of the cloud migration challenges
- Investment to drive the migration efforts
- Mechanisms to evaluate the migration's success

Lack of data security and risk assessment

The minute you start transitioning data from your **Data Centers (DCs)** to the cloud without a proper plan, your data becomes vulnerable to potential security risks. Many enterprises do not want to run their applications on the cloud, assuming that they will risk their companies' data security and compliance. They fear that their data could be exposed to cyber-attacks, leading to serious disruptions during the migration process.

Companies should make decisions by thoroughly vetting the cloud vendors and the compliance programs that they have completed. Understanding how secure the cloud provider's data centers are, along with the list of certifications from accreditation bodies, will ensure that data is not visible to anyone except you, thereby eliminating the risk of any kind of data theft during the migration.

Focus on checking the following before your migration:

- Identifying vulnerabilities, both short-term and long-term, that the migration can cause
- Monitoring your workloads for threats and generating alerts for them in real time
- Encrypting and backing up your data end to end
- Aligning the cloud provider's security policies with that of your workload
- Compliance standards such as the HIPAA, PCI DSS, CCPA, and so on, which you need for your data

Lack of accurate migration budget assessment

Cloud migration costs can sporadically increase as a result of all the challenges discussed earlier. This leaves enterprises spending more than the actual estimated costs, which could potentially lead to revenue losses.

It is important for companies to accurately factor in operational and administrative costs, and scope the expenditure to achieve a manageable migration. Moving forward with cloud migration without proper analysis can result in unexpected budgets, leading to hindrances to the project.

Considering the following is advisable:

- Expenses to cover hardware and software licenses
- Administrative expenses to cover staff training, system management, payroll, and so on
- Operational expenses

Lack of on-premises-to-cloud compatibility

Many organizations face the challenges of their applications not being in a migration-ready state or compatible with the cloud framework. Identifying such interoperability issues and addressing them upfront through tweaking or extensive redesigning if needed is an important decision to make.

As we looked into the challenges that companies face when it comes to cloud migration, and the various dimensions that we need to keep in mind, let's take a look at the essentials that enterprises need to incorporate for successful cloud migration.

The evolving benefits of cloud migration

While the benefits of the cloud are emerging and evolving day by day, the key driver of cloud migration is almost always its perception as a cost-cutting measure. Despite the cloud's ever-growing capabilities, many businesses look at the short-term benefits rather than their long-term goals.

As your adoption grows, redefining your transformation strategy is required for optimizing your workloads that are deployed on the cloud. Modernizing your applications, empowering your staff, and transforming your workloads will have a *long-lasting impact* on the way you are operating on the cloud. Many organizations that are **born in the cloud** often witness freedom from technical debt and overhead, where they are able to carve out a niche place in the competitive market offering innovative and customer-focused business models.

Let's take a look at what building a cloud-first mindset involves and why is it important in the next section.

Building a cloud-first mindset

Enterprises across every industry that are yet to transition to the cloud are losing a competitive edge against their peers who are running their applications with a cloud-first strategy. Deploying workloads on the cloud can eliminate technical debt as well as accelerate your digital transformation.

It requires a change in an organization's culture to achieve the cloud-first mindset for effective cloud-native adoption. It is important to rethink your way of doing things and understand that the cloud may be the new normal for new projects and initiatives.

With the recent *COVID-19 pandemic*, there have been worldwide disruptions that have pushed organizations to revisit/rethink their **business continuity plan** strategies. Many businesses are facing tremendous challenges to maintain their infrastructure in **DCs**, often resulting in a costly effort.

Organizations are realizing that the quick adoption of cloud technology can help them to address these challenges. To achieve results in this highly competitive business landscape, businesses are strategically approaching cloud migration versus doing so tactically, which is the biggest evolution in this cloud era. Systematically thinking about the following three aspects can help you look at your migration efforts more from a business perspective than just a technical effort.

Assess – understanding the present

Ensuring that your IT landscape is well suited to the cloud and knowing whether new design or architecture is required is an important conversation to have as a CTO early in your organization's migration journey. Having essential insights into your current application landscape – its architecture, usage, resources, and dependencies at a 10,000-feet overview – will help get an accurate assessment.

Vision – imagining the future state

Having a vision statement provides both technical and non-technical stakeholders with a guiding light and a purpose that your organization hopes to achieve through cloud migration.

As an organizational leader, you need to ask yourself what it will look like when you migrate to the cloud and how impactful it will be for your business and its customers. This is where you can channel the utmost creativity and envision the future to drive the team and inspire them.

Mission – setting the goals

Setting your goals and objectives for migrating to the cloud is where you will begin to define specific and measurable outcomes. Many aspects, such as your workforce productivity, infrastructure costs, latency, and time-to-market delivery, play an important role in migrating your workloads to the cloud. Defining how each of these should look is a decision that will enable your migration model.

The *assess, vision, and mission* model is a smart route when it comes to looking at your workloads and incorporating a cloud-first mindset. With its great efficiency, you can become a critical enabler for your company's accelerated innovation and advance your digital technologies. Let's look at the various phases of cloud migration to maximize its value.

Exploring the phases of cloud migration

Cloud migration is a business-driven decision. Migrating to a cloud platform usually involves end-to-end business transformation, not just technical, and can be well represented by five phases, namely **Discover**, **Plan**, **Migrate**, **Automate, and Optimize**. The following phases present a framework that is key to a successful migration.

Phase 1 – Discover

The first and foundational phase for the rest of the phases in cloud migration is the *Discover* phase. Organizations not investing in this phase are bound to fail, especially if an organization affects many teams and touches applications with complex dependencies.

Understanding aspects such as the technologies used by the current workload's landscape, compliance requirements, dependencies within the applications, infrastructure needs, network, and cost structure helps you to identify relationships that cannot be found in any application implementation documentation. This context can then be used for mapping relationships between your organization's hardware, software, security, and operations.

Here are some of the recommendations that you can incorporate to become successful in this phase:

- **Assessing cloud readiness**: Many organizations that are planning large-scale migrations to the cloud are seeing the need to deploy **automated discovery tools** and accelerate the discovery step. The tools do the heavy lifting when it comes to performing extensive discovery of an organization's IT portfolio and collecting the right candidates for the cloud migration, including compute, network, database, and storage workloads.

  IT portfolio discovery is a crucial element of the discovery phase, where a detailed enterprise-level knowledge of the legacy applications along with the list of dependencies can be understood from this exercise. The decision-makers will be able to get an accurate list of when and how to migrate workloads through this effort.

- **Dependency mapping**: Dependency mapping is the process of discovering and documenting the interdependencies between various applications and systems. Identifying the dependencies is a step that is often time-consuming and complex, especially for legacy applications or systems that have been running for years.

  For example, a credit card application service may be using databases, external **application programming interfaces** (**APIs**), and other internal shared services that are accessed via the APIs. The teams responsible for dependency mapping will identify and document these dependencies.

- **Interviews**: Manually capturing data about workloads through interviewing business owners can help fill in undocumented aspects or even verify some of the documented learning about the workloads that are running as legacy applications.

- **Consolidating and documenting:** As you iterate through the preceding steps, ensure that you are consolidating and documenting the relevant details, diagrams, references to contracts, data center agreements, databases, and compliance information. This information not only serves as the reference for your current cloud migration discovery efforts but also serves as the audit reference and single repository of truth for your organization.

Phase 2 – Plan

Once you have the Discover phase in place, you can start planning out your migration process by defining the list of workloads and infrastructure that will be part of the migration. You use the mapping that you captured from the discovery phase and categorize the applications and dependencies that will be moving to the cloud.

In addition to the workload scoping, having a clear understanding of whether you are migrating these workloads to a single cloud, multiple cloud providers, or a hybrid cloud is important, as the effort will vary depending on what you choose. Scoping these details out will enable you to assess your readiness and help you gain confidence by starting with workloads that are ready to be migrated.

Typically, the single-cloud approach is the simpler approach given that you are focused on one cloud provider's offerings versus a multi-cloud approach, which can get more complex. Although with the multi-cloud approach, you may achieve greater strengths and advantages of each cloud, that again depends on what your final state looks like. With a hybrid-cloud approach, you get greater flexibility and agility with your deployment and connectivity options, although you must lay the extra groundwork on how you would likely want to move your workloads between the private and public environments. It is also important to take the complexity of your workload into consideration, as that can make or break your migration journey along with the aforementioned approaches.

Along with the deployment model, you determine whether the service category is **Software as a Service (SaaS)**, **Platform as a Service (PaaS)**, or **Infrastructure as a Service (IaaS)**.

This phase is also the perfect time for you to make adjustments and reiterate your strategy document. As you dig deeper and identify any changes that you may need to support the transition successfully, discuss them with the stakeholders and document how to train your existing teams to bridge their knowledge gaps.

Phase 3 – Migrate

Next is the Migrate phase, where the actual heavy lifting happens, and you start with executing the migration. Every organization has its own set of protocols, budget constraints, culture, and go-to-market timelines that influences its decision-making process. Typically, starting with low complexity will give you the experience to carry over migrations for high-complex applications.

Cloud migration process checklist

There are many aspects that an organization needs to prepare for to achieve a successful cloud migration.

Here are some of the industry-proven checklist items of a cloud migration process. We will be diving deeper into each of these aspects in the next two chapters:

- Did you define the business goals for the migration?
- Did you choose a cloud environment (single, hybrid, or multi-cloud)?
- Did you evaluate your migration budget?
- Did you choose a cloud provider that fits your long-term needs?
- Do you have the right architecture to migrate?

- Do you have a plan to test migrated applications, databases, and their dependencies?

- Do you have a plan for prioritized workloads to be migrated?

As you collaborate and come up with answers for each of these checklist items, you will be better equipped with ways to implement and manage your migration to the cloud effectively.

Phase 4 – Automate

Organizations must continuously evolve to achieve cloud migration. And the one factor that enables them with agility is **cloud migration automation**.

Cloud automation refers to automatically installing, configuring, provisioning, and managing cloud resources and their operations using tools to reduce the manual burden of doing so.

Automating cloud migration holds significance due to many reasons, such as the following:

- **Reduces timelines**: Traditionally, migrating to the cloud involves many phases such as the design, development, and deployment of infrastructure on the cloud, which involve manual processes. Incorporating tools to automate this can reduce time and human effort significantly, while not compromising the structure or integrity of your workloads.

- **Reduces misconfigurations**: Human errors are typically the most common cause of misconfigurations that lead to a compromised infrastructure. By automating the repetitive tasks, there is no room for errors in the process.

- **Lowers costs**: Cloud migration entails many repetitive task, which are often performed by IT staff, and with automation, organizations can benefit from reducing the costs without burdening their staff. This can help you to accelerate your migration process by applying your staff to many meaningful tasks and expediting your migration process.

- **Minimizes business downtimes**: Organizations that implement mass migrations to accelerate their cloud adoption journeys often need to move large volumes of data without any disruptions to their businesses. Using automation can help them reduce longer cycles and minimize any impact on business operations.

- **Supports compliant data conversion**: For organizations that require data to be stored in specific ways for security or compliance reasons, the data conversion process is required before migrating to the cloud, which is often time-consuming and costly and results in inconsistent processes. Automating data conversion can help cut down these long implementation cycles and enhance the data conversion efforts significantly.

Automation tool evaluation checklist

There are many vendors that have automation tools readily available for your migration efforts, but choosing the right tool is the key decision that you will make to drive your efforts successfully.

Here are a few factors to remember:

- Does the tool support multiple cloud platforms such as AWS, Azure, Google, Oracle, and so on?
- Does the tool focus on security?
- Does the tool focus on compliance?
- Does the tool support resource backup and disaster recovery (DR)?
- Does the tool support service enablement and disablement?
- Does the tool help with inventory tagging?
- Does the tool provide support for provisioning and auto-scaling?
- Does the tool focus on cost?
- Does the tool focus on performance?
- Does the tool allow you to take backups at specified intervals?

Phase 5 – Optimize

This phase is as important as any other, if not more so, to ensure you are achieving your long-term business goals. Running your workloads efficiently and reliably is dependent on how well you plan to optimize while you operate. This phase is a crucial factor for any enterprise to achieve its goal of getting the best out of the cloud. In the next few chapters, we will talk about how to iteratively optimize your workloads that are running on the cloud.

Finding new ways of optimization should be an ongoing and continuous task, where you constantly remove uncontrolled and redundant tasks and define your optimization requirement goals. Once you deploy your workloads in the cloud, you refine them and fine-tune their infrastructure to make them more efficient. With optimizing activities, you achieve goals centered around areas concerning the savings, performance, security, and resiliency of your application stack.

Operational efficiency can never be achieved overnight. Each workload presents incremental learning to improve on optimizations, and as you incorporate them into your existing workloads, you will understand and master the process of planning for your future migrations and how best to maximize the benefits of these platform improvements.

Here is an optimization checklist that you can make use of to better define your optimization phase:

- Do you have a monitoring system for your workloads to measure your environment?
- Can you analyze data coming from these measurements?
- Do you have a plan to roll back changes without manual intervention?
- Can your workloads scale vertically or horizontally?

- Are you meeting the operational cost goals of your environment?

- Are you able to increase the innovation velocity of your team?

- Are you able to reduce time-to-market?

- Do you have an incident management runbook for your infrastructure?

- Do you have **Recovery Time Objectives** (**RTOs**) and **Recovery Point Objectives** (**RPOs**) defined for your workloads?

- Do you have a DevSecFinOps team culture?

Understanding and iterating through the processes of optimizing can help you look for bottlenecks and define your future optimization goals. Don't stop your checklist here; build on top of this based on how you would effectively manage a cloud-native environment and operate your workloads in a way that is scalable, resilient, manageable, and observable by services without having to set up a complex solution.

Summary

In this chapter, we learned how any enterprise can balance its business drivers and technology needs to achieve the right blend of speed and scale with cloud migrations. We also looked at the key factors to consider in the migration journey to ensure successful and maximized benefits. Organizational behavior and mindset are key when it comes to the transformation of their IT portfolio, and identifying roadblocks in a timely manner will change the landscape of your migrations. The next chapter will focus on the key aspects of cloud migration while preparing for it.

Further reading

- https://www.gartner.com/doc/reprints?id=1-271OE4VR&ct=210802&st=sb

- Summary of the HIPAA Security Rule - https://www.hhs.gov/hipaa/for-professionals/security/laws-regulations/index.html

- PCI Compliance Guide - https://www.pcicomplianceguide.org/

- AWS Cloud Essentials GETTING STARTED GUIDE - https://aws.amazon.com/getting-started/fundamentals-core-concepts/

- Introducing FinOps—Excuse Me, DevSecFinBizOps - https://aws.amazon.com/blogs/enterprise-strategy/introducing-finops-excuse-me-devsecfinbizops/

- Infrastructure as code - https://en.wikipedia.org/wiki/Infrastructure_as_code

- AWS resource and property types reference - `https://docs.aws.amazon.com/AWSCloudFormation/latest/UserGuide/aws-template-resource-type-ref.html`

- API Reference - `https://docs.aws.amazon.com/cdk/api/v1/docs/aws-construct-library.html`

- Operational Excellence Pillar - AWS Well-Architected Framework - `https://docs.aws.amazon.com/wellarchitected/latest/operational-excellence-pillar/welcome.html?e=gs2020&p=fundcore`

- What is shuffle sharding? - `https://aws.amazon.com/builders-library/workload-isolation-using-shuffle-sharding/?did=ba_card&trk=ba_card#What_is_shuffle_sharding.3F`

3

Preparing for Cloud Migration

In the previous chapter, we discussed why it is important to have a cloud-first mindset and how organizations are thinking of moving to the cloud amid the growing challenges of worldwide disruptions. Both business and technical teams need to accept cultural changes and support each other's teams through the migration journey.

This chapter will prepare you to succeed in cloud migration by looking at a cloud migration strategy as a systematic and phased approach but not as a tactical technical undertaking. The cloud migration services market is growing rapidly, and it is important to know the key considerations when choosing the right partner for cloud migration. We will discuss in detail various things to be aware of as you make these choices in this chapter.

In this chapter, we are going to cover the following topics:

- Learning about cloud migration insights
- Choosing the right cloud partner for your business
- Unraveling the multi-cloud—benefits, challenges, and strategy
- Aligning your **information technology** (IT) landscape with the cloud—best practices

Learning about cloud migration insights

It is easy to overlook the essentials of cloud migration when you are deep into the weeds of it. There is a broad range of migration considerations that each organization needs to make while working closely with the stakeholders.

Let's look at some key insights of cloud migration that will apply to every organization. As you make cloud choices throughout your journey, you will find these learnings will guide your cloud strategy.

Begin with cost savings, evolve with innovation

The first factor that drives many organizations to the cloud is cost reduction. Reducing operational costs for smaller start-ups or large organizations is a key aspect for them to invest in the elastic model of a cloud platform. Focusing on an incremental approach enables many companies to expand their evolution to domains such as security, performance, and innovation. Businesses are no longer trapped in a data center and its software and can take ideas from inception to reality much faster.

Cloud migration propels cultural shift

The cloud not only changes the way your devices and data interact with one another but brings a cultural transformation that affects how people work together, in a powerful way.

Understanding how the cloud will impact your organizational culture

Many companies running legacy applications will have to let go of a siloed culture and begin to share via the cloud. This could itself result in a massive cultural change, shattering the barriers of a closed corporate culture. Creating a new culture where there is information sharing and decentralization will pave the way for a new operational paradigm.

Getting leadership onboarded to act as change advocates

One constant for businesses of all sizes across the globe that can lead to a successful business transformation is solid leadership with executive sponsorship. Making high-level decisions, providing resources to the initiative, and making sure teams are adopting across the company as a priority are critical aspects to the success of any change. Driving this change consistently to insist on and sustain the momentum and ensuring that the process works across all departments is also important when trying to bring change to the organization.

Bringing awareness of this change through communication

A challenge that keeps coming up is the first sign of resistance to a cultural shift. This usually arises as a natural response from staff when they feel that a sense of ownership is lost. For example, if an emerging cloud culture involves gaining knowledge of new technologies and the company doesn't make any resources available, it will hamper staff confidence and could result in fear and negativity.

You need your employees to be on board with these changes, and they often accept change when this is communicated effectively, helping them understand the benefits of doing so and involving them in the process early on. Frequent communication and inclusion are the top drivers during times of change, and leaders can receive unique insights from their employees, especially **subject-matter experts** (**SMEs**) when they are involved in decision-making and during times of change.

Recognizing and addressing any concerns up front

An understanding and willingness to look beyond surface issues and address them has a high potential to drive the culture shift successfully. For example, an organization that handles secure data raised concerns throughout the company's departments when they first came to know about the company's cloud strategy and expansion of access to their data. But when leadership demonstrated the measures and showed that they plan to implement cloud technology with the utmost attention to security, these insights let the focus shift from leadership to the teams themselves.

The preceding two aspects are important for any organization leaders to incorporate early in their cloud journey to ensure a successful transition.

As thought leaders, here are four areas that need to be addressed in order to bring about this change.

High-level metrics are not enough

Capturing metrics while running on the cloud is different than on-premises. The cloud requires the tracking of more metrics as there are additional considerations that are needed for the collection and analysis of these. Metrics such as response time, requests per minute, error rates, average compute cost, and average storage cost are all key metrics when it comes to cloud environments. To account for factors such as distributed architectures, shared ownership, cost, latency, and load balancing, you must look at metrics in depth in addition to **infrastructure consumption metrics** such as **central processing unit** (**CPU**) usage, disk **input/output** (**I/O**), and so on.

Proper planning drives successful migration

Cloud migration is one of the biggest transformations that organizations will experience. For an effective cloud migration, it is critical to plan your migration rigorously. Based on learnings from many complex and large migration projects, it is important to have a **plan for migration execution** as well as a **rollback strategy**.

As you start gaining some experience, you will realize that building a business case, assessing your current IT landscape, and building a migration plan will give you an **end-to-end** (**E2E**) view of efforts and influence key decisions that you will be making.

Thinking beyond CapEx and OpEx

An organization makes a significant **capital expenditure** (**CapEx**) commitment to procure hardware, software, and infrastructure upfront in traditional data centers. Even though you have full control of your resources, you run into other constraints such as maintenance overhead and having to own or rent physical space. With the cloud, you will notice a shift of finances from CapEx to **operational expenditure** (**OpEx**), where resources are decentralized, and you have easier access to a pool of resources as your demand varies.

Conducting **total cost of ownership** (**TCO**) and **return on investment** (**ROI**) will give you the ability to define goals that you want to achieve as you run applications in the cloud. With the cloud, you can look beyond cost savings and think about the wide variety of reasons a given application of your business will benefit. This will help you to ensure that you have a solid migration plan from the beginning and achieve a winning cloud-migration game plan.

There are many great cloud providers

Many companies are coming up with cloud services that offer **software as a service** (**SaaS**), **platform as a service** (**PaaS**), and **infrastructure as a service** (**IaaS**). When you consider how cloud computing has evolved over the last decade, you will see how cloud-native has emerged and become established as the foundation of IT. Companies realize that the shift is an **absolute business necessity**, and given that cloud management platforms have sprung up offering hundreds of solutions, the shift is drastic.

Choosing a cloud partner that is most suitable for your business and requirements is no doubt a tedious task. We will be diving deeper into the types of cloud providers and some considerations that you need to keep in mind while evaluating a cloud provider.

Choosing the right cloud partner for your business

To achieve long-term success, partnering with the right cloud provider is the key factor. This, in conjunction with an understanding of your workloads, will give you an assessment of how much work is needed to be able to migrate to the cloud.

The available market for cloud providers is vast. Providers such as **Amazon Web Services** (**AWS**), **Google Cloud Platform** (**GCP**), and **Microsoft Azure** have garnered the largest cloud market share, while there are many other niche players such as **Rackspace** and **IBM Cloud**, offering more specialized services.

In this section, we are going to discuss key distinctive factors that you can use to evaluate and identify a provider that will support your unique business, technical, and operational characteristics in an optimal way.

Security

Security is the foremost criterion and an area of priority in the cloud. Have a clear understanding of what your security goals are and map them to the provider's security measures. In addition, evaluate the cloud's **shared responsibility model** to understand features that are provided for free and features that have an additional fee. Considering third-party partner technology to supplement missing features is also a general practice.

The following diagram illustrates the cloud's shared responsibility model that customers and cloud providers are generally responsible for:

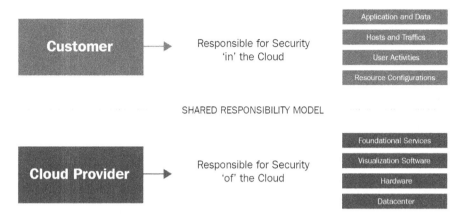

Figure 3.1 – Cloud shared responsibility model

The cloud provider is responsible for the **security of the cloud**, including hardware, software, networking, and infrastructure that runs cloud services, while the cloud provider's consumer is responsible for **security in the cloud**, including applications, their configurations, and their usage.

In general, any user action and activity should be auditable and should have structured processes and mechanisms in place for roles and responsibilities. Ensure that the cloud provider's maturity of security operations and governance processes is assessed thoroughly and any risks are laid out clearly in your decision-making process.

Certifications, compliance, standards, and attestation frameworks

Certifications can be an important aspect when shortlisting potential cloud providers. Cloud providers try to adhere to industry best standards by complying with recognized certifications and standard frameworks. Certifications such as **International Organization for Standardization (ISO)** *27001*, the **Cloud Computing Compliance Criteria Catalogue (C5)**, and the **Federal Information Processing Standards (FIPS)** are assessed and certified by third-party and independent auditors.

The following screenshot illustrates some common standards and certifications:

Standards organisations and frameworks - examples;

Cloud		Security		Operations	
DMTF	ISO	CSA	ISO	ITIL	ISO
ETSI	GICTF	SSAE 16	PCI	IFPUG	CIF CERTIFIED
OpenGridForum	SNIA	GDPR	IEC	DMTF	COBIT
OPEN CLOUD CONSORTIUM	Cloud Standards Customer Council	COBIT	ICO.	TOGAF 9	MOF
NIST	OASIS	CYBER ESSENTIALS	HIPAA	tmforum	
IEEE	I E T F	ISAE 3402		FitSM	

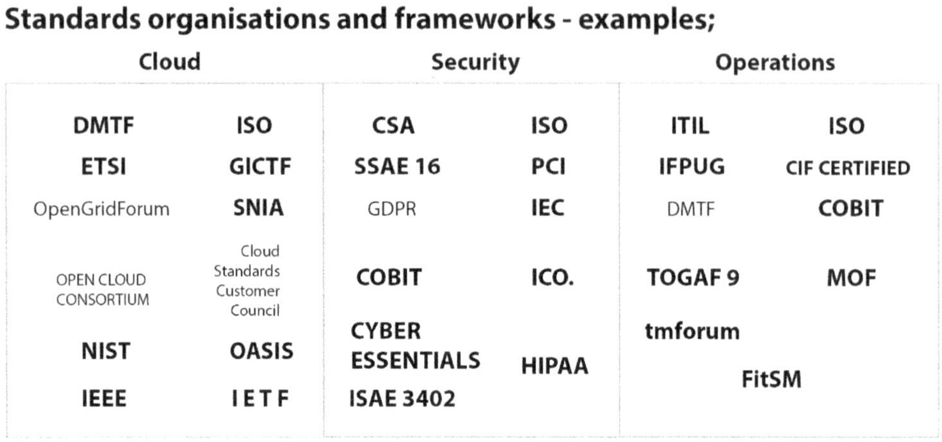

Figure 3.2 – Common standards and certifications

IT standards are broken down into *certifications and attestations*, *laws*, *regulations*, and *privacy*. There are also certifications that are broken down for specific industries, such as educational institutions, government agencies, financial services, and healthcare providers.

Look for standards and processes that are important to your organization. Understanding how cloud providers plan to work toward continuous adherence to standards that are important to you is a good evaluation exercise.

Innovation and roadmap

The continuous innovation of a **service provider** (**SP**) is important for your organization's agility while running in the cloud. For those who use the cloud in their daily operations, cloud technology allows businesses to streamline and function effectively. For many, the possibilities that the cloud offers, along with limitless technology such as computing and data storage services, help businesses to grow, innovate, and integrate with their partners or customers seamlessly.

You should consider choosing a cloud provider that has a fast pace of innovation and the broadest and deepest functionality that is important to your applications. Access to such a flexible external platform helps companies be more productive and innovative and scale faster.

Ask a provider how they plan to evolve and develop their services to see if their roadmap fits your needs. Some principal factors to consider are their investment and a strategic roadmap to specific technologies, as well as the ongoing commitment they can offer. Depending on this, you also want to evaluate their overall portfolio and how interoperability and integration with SaaS providers—if applicable—will play over time. In general, it is important for SPs to provide a good variety of compatible offerings.

An ever-growing feature set that will give you the required resources and services to build is what is needed for today and tomorrow, while keeping your data secure, saving money, and producing better products can help businesses get closer to customers, employees, and partners to develop efficient communities.

Service-level agreements and contracts

A **service-level agreement** (**SLA**) outlines what a **cloud SP** (**CSP**) is responsible for when offering its services. The contract covers responsibilities that the CSP guarantees binding in an agreement. Some of them are **service-level objectives** (**SLOs**), data policies and protection, cloud infrastructure, customer responsibilities, **disaster recovery** (**DR**), and backup.

An SLO is an important aspect that covers service availability, penalties, and service capacity. Let's look at each of these points one by one, as follows:

- **Service availability** is quantified in terms of a time frame as a **quality-of-service** (**QoS**) concept. For example, a cloud provider may offer 99.99% availability, including alerting its customers and providing maintenance updates or service downtimes. This breakdown of uptime is helpful to know upfront to assess if your business can afford the unexpected downtime. Look for a clear measurement of service availability. Essentially, you should have an understanding of how responsibility is distributed between the customer and the provider in aspects such as service delivery, provisioning, monitoring, and so on.

- **Penalties/compensations** are service credits related to service downtime that are offered by the cloud provider to the end user. Monitoring and logging mechanisms come in handy for claiming, as well as for calculating service credits. Ask for examples of scenarios and compare differences in penalties across multiple providers.

- **Service capacity** is a balancing act that every cloud provider establishes to avoid overprovisioning of users, connections, and resources. CSPs are faced with the task of estimating capacity in order to meet their SLOs.

- **Data policies and protection** establish who owns data in the cloud. When you have sensitive data, this is a critical aspect that will help you determine whether you want to make the transition to the cloud. For aspects related to data privacy regulations, ensure that there is clarity and sufficient ground to establish guarantees around the access of data, its location, and confidentiality on usage and ownership rights. Transfer of data and its conversion if you decide to leave the organization should also be reviewed.

An SLA defines a CSP's data ownership policy clearly. Typically, ownership rights are on the consumer, and you should ensure that ownership is explicitly stated on the contract to avoid any confusion.

Cloud infrastructure is the hardware and software that the cloud provider requires to run and operate its services. The provider should outline at a high level details of the hardware that services rely on. This will help you understand the robustness of the CSP's infrastructure to make an informed decision.

DR lets you have a plan to prevent downtime of your applications running in the cloud. **Backup** of data and applications will enable you to plan for the avoidance of any loss of your data. Understand the cloud provider's ability to support your **recovery time objectives** (**RTOs**) in terms of data sources, scheduling, and backup and restore processes.

For example, *automatic backups and snapshots* is a feature that CSPs should provide so that users can set it up and have a solution to activate when required.

Customer roles and responsibilities are outlined in the SLA, and both the provider and the customer will need to agree on these. Understanding and staying informed on what you are liable for as an end user is important before you enter into an agreement, and this is always a task that you need to include in your checklist of items to evaluate.

ISO/**International Electrotechnical Commission** (**IEC**) *Cloud computing – Service level agreement* is a framework that establishes common terminology and defines a model for specifying metrics for cloud SLAs. This is a useful framework to use when assessing a cloud provider's agreement. It is a good idea to challenge cloud providers to offer flexible terms on how they plan to support these aspects.

Support model

When organizations of any size plan to move to the cloud, it is undoubtedly a transformation journey that their teams need to be prepared for. The transformation involves not just a technology shift but also a shift in business processes, including usability improvements.

Cloud transformation is a journey that begins with implementation. Through this journey, *day-to-day support* of these processes and functions is crucial in order to reap the benefits of the cloud. Support after implementation is a crucial element to maximize the benefits of cloud applications.

Given that the nature of on-premises and cloud is different, the type of support is also different. Let's look at the following table to get a comparative view of support for workloads running on-premises and in the cloud:

Parameters	On-premises	Cloud
Infrastructure	Applications running on-premises need substantial infrastructure support	Infrastructure in the cloud is managed by the cloud provider, so you do not require infrastructure support.
Support team skillset	Technical	Technical and functional expertise related to services that you leverage. Design principles, best practices, and architectural choices can be discussed with the support teams.

Parameters	On-premises	Cloud
Support activities	• Patching for vendor bugs • Enhancement of hardware • Updates to product functionality • Rollout support for enhancements	• Functionality guidance support • Rollout support for enhancements
Support duration	Dedicated support	On-demand support

Table 3.1 – Support for workloads running on-premises and in the cloud

Many organizations end up spending millions to set up on-premises IT support for their applications and infrastructure. This overhead is changing significantly as you see a shift to the cloud.

When running in the cloud, there is a list of activities that you require your CSP to offer. This usually comes as a subscription service where CSPs have a framework of dedicated support teams handling activities such as the following:

- **Technical support**, through which you get access to technical experts who can help with queries and support to help deploy, operate, or optimize the cloud environment based on your use cases.

- **New features support** is important when cloud vendors release new features and enhancements through release updates. When these updates are launched, end users look for assistance to enable smooth operations of the service. Through this support, specialists can be available to help customers by providing the required guidance so that applications run without any hassle.

- **Billing support**, to provide guidance for any billing-related questions, such as usage of services, account-related billing, or general cost management.

- **Migration support** provides access to support when customers are executing migrations such as application launches or infrastructure migrations, and even for large-scale events such as live streaming or marketing events that require scaling. Ensuring that cloud providers can help with additional planning and providing experts to review and identify any risks for such large events is a value-add and a key feature that you should be looking for within any support provided.

- **Functional support** provides access to SMEs who can help you manage the functional aspects of services and enable you to build optimal processes.

- **Partner integration support** for when you are using third-party SaaS applications in conjunction with cloud providers' services. Many CSPs realize that their services cannot operate in isolation. As they evolve, they need to be able to integrate them with SaaS applications to address the customer landscape of use cases. The support provided by CSP cloud experts can go a long way in such cases.

Consider the details mentioned here and discuss with the provider how best they can accommodate the mentioned support options.

Pricing model

Every cloud provider offers a unique package of services and pricing models. The differentiating factors are typically dependent on the usage period, allowing granularities for per-minute usage, as well as pricing agreements or discounts for long-term commitments.

When it comes to variations of the pricing model, there are three types—consumption, package/ subscription, and configurable, as shown in the following screenshot:

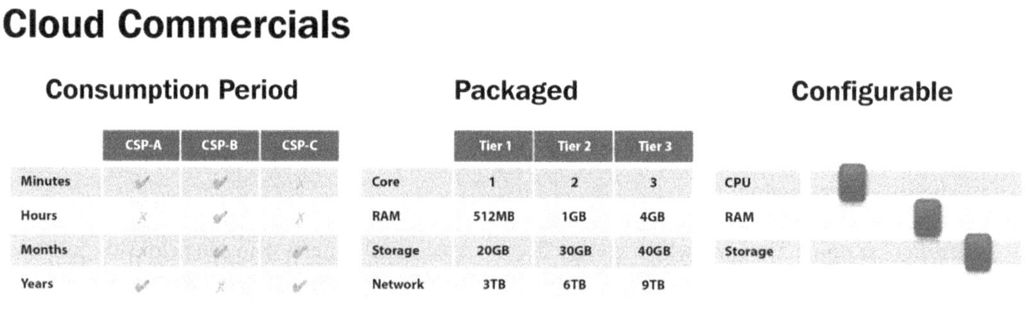

Cloud Commercials

Consumption Period	CSP-A	CSP-B	CSP-C
Minutes	✓	✓	✗
Hours	✗	✓	✗
Months	✗	✓	✓
Years	✓	✗	✓

Packaged	Tier 1	Tier 2	Tier 3
Core	1	2	3
RAM	512MB	1GB	4GB
Storage	20GB	30GB	40GB
Network	3TB	6TB	9TB

Configurable	
CPU	
RAM	
Storage	

Figure 3.3 – Variations of the pricing model

The **consumption** (**pay-as-you-go**) model is where the consumer starts off with a zero account balance and gets charged based on resources provisioned on demand.

The **subscription** or **packaged** model, whereby services are sold on a catalog basis, is billed per month or per license, and the consumer gets billed for those resources, whether used or not.

The **configurable** or **hybrid** model (pay-per-use plus subscription) is enterprise-billed in advance based on the number of active resources and dedicated resources for a period.

In general, the pricing model in the cloud is more flexible than the traditional model, and the value chain is usually driven by economies of scale. Factors such as monitoring and the cost of data centers can influence the pricing of cloud resources. Each CSP's pricing model comes with its own advantages and disadvantages, so ensuring the framework is favorable to you is a major consideration to make.

Vendor lock-in criteria and exit provisions

Customers often end up depending on specific SaaS applications offered by a provider and implement workloads that cannot be easily moved in the future without substantial changes. This lack of standardization results in **vendor lock-in**, whereby businesses are stuck with a particular service.

Technical limitations arise because of tightly coupled services that are tied to cloud platforms. Devices and software working only when associated with a specific vendor lead to interruptions in operations and businesses.

An inability to move applications or data across different cloud platforms can cause interoperability issues, which can often result in tedious and costly tasks. This causes *an unsustainable status quo* where you may be losing money and staff.

Avoiding such risks of vendor lock-ins by choosing providers that have technology you can easily port to other cloud providers is a key criterion you want to look for. Select services that have comparable alternatives in the CSP market to *minimize* lock-in risk. Consider reviewing these options periodically to be cautious of feature parity, whereby service platforms may keep introducing new lock-in factors through update launches.

There could be a few compelling benefits that are offered by key providers, but assessing them carefully and balancing these benefits so that you do not get dependent on a specific provider is an ideal strategy. It is always a good idea to retain the flexibility of changing providers as per your business needs. Let's look at some strategies to avoid vendor lock-ins.

Understanding infrastructure and dependencies

To start with, perform a thorough audit of your on-premises infrastructure, including hardware and software. This will give you a good understanding of the business and operational requirements. The components of such workloads that are dependent on legacy technologies or infrastructure may not leverage solutions from an external vendor. Through this exercise, you will be able to identify legacy workloads that are designed to operate on-premises and that you are likely going to benefit from by operating them in this way.

Understanding common features

Once you review the existing technical stack, do a comparison of what is compatible across existing cloud vendors. When you identify such common functionalities, you can determine the best cloud solution for your needs. Your decision to migrate and the technical requirements all depend upon whether an alternative service is available widely in the market or not.

Redesigning or upgrading before migrating

As we discussed in the earlier criteria, incompatible workloads that are dependent on a limited set of legacy technologies may also warrant an upgrade or rework before migrating to the cloud. Evaluate the best option considering your future requirements, such as scaling to expand the user base, before determining the best course of action.

Maximizing the portability of data and applications

Data portability improves the ability to migrate or move data from one computing platform to another easily. In order to achieve data portability, vendors should define their models in a way that increases usability across platforms. It is becoming increasingly important to store data in a portable format when it comes to moving data to the cloud.

Although avoiding vendor lock-in is one of the reasons to implement data portability, factors such as improving resiliency and reducing costs on storage tiers and compute platforms drive the need further along. The following screenshot illustrates the importance of data portability:

Resiliency	**Avoiding vendor lock-in**	**Cost**	**Exploiting cheap compute**
If data can move among cloud providers, it remains accessible even if one cloud has an issue.	Mobile data means customers don't have to commit to a single provider because it offers better data services.	Mobility lets customers take advantage of cheaper tiers of storage from among various cloud service providers.	Portable data lets customers take advantage of discounts compute providers offer at certain times of day.

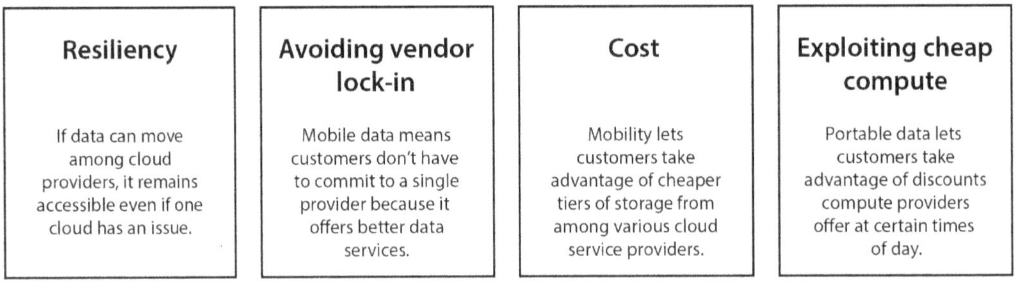

Figure 3.4 – Importance of data portability

Different cloud providers may have data formats that can lock you into specific platforms.

Technology standard frameworks such as the **Cloud Data Management Interface** (**CDMI**) should be followed to enhance the interoperability of storage systems. The CDMI framework defines how applications should be creating, retrieving, updating, or deleting data from the cloud.

Another commonly used framework is the **Open Data Element Framework** (**O-DEF**), which standardizes how basic units of data are classified. This simplifies software development and improves the management and organization of data overall.

Open data formats make data portability easier and improve the standardization of formats given the support and documentation that is available. **JavaScript Object Notation** (**JSON**), **Extensible Markup Language** (**XML**), and **comma-separated values** (**CSV**) are some examples of interoperable formats.

In addition, here are some best practices to maximize data portability across your organization:

- Ensure that your teams can recognize requests that are data-portable
- Ensure that your teams know which types of data have the right portability for reusing
- Ensure data is transmitted in a structured machine-readable format that is commonly used
- Transmit data in a secure way

Application portability

Application portability ensures that you are building applications that can be supported by multiple cloud platforms. Applications that are running on IaaS or PaaS deployments should be designed in such a way that they are easily decoupled from the underlying infrastructure or platform. This is usually possible when the applications are built leveraging **REpresentational State Transfer** (**REST**) **application programming interfaces** (**APIs**) with popular open data formats such as **HyperText Transfer Protocol** (**HTTP**) and JSON, which we discussed earlier. Vendors should define ways to abstract business logic from the underlying infrastructure and enable data transfer mechanisms when required.

If your workloads are designed with the anti-patterns that we discussed or if the vendor is not providing an environment where best practices are not incorporated, you know that vendor lock-in is likely. Failing to adhere to open standards is also an anti-pattern that you want to look out for.

Considering a multi-cloud strategy

Enterprises often express concerns about losing control of business-critical data and applications when they depend on a single supplier. Uptime and security are their top two causes for concern. While it is natural to feel so, businesses don't have to feel limited when it comes to cloud providers. A well-thought-out multi-cloud strategy can meet modern business requirements as well as let you have peace of mind.

We will be discussing multi-cloud pros and cons in detail, but initiating a multi-cloud strategy can help you spread across multiple clouds to avoid business lock-in and feeling trapped in a specific platform.

Planning an exit strategy

Before you sign off on a contract, ensure that you plan an exit strategy and negotiate exit provisions when the time comes to leave. Ensure that you have a **clear exit strategy** in place. It is recommended that leaders understand the terms and conditions of service-termination clauses provided by the vendor. Tracking all the costs of exit, such as storage, data and third-party costs or maintenance of licenses, is an important task to spend time on.

Reliability and performance

There are many ways to evaluate and measure the reliability of a CSP. Ask how many times the SP had service degradations in the last 12 months. Some vendors are very transparent in sharing such information, and some should be able to provide you with this when asked. Keep in mind that there will be system failures, and downtime is bound to occur. The important aspect is how the provider deals with that downtime.

Verify that the provider has monitoring and reporting tools that can integrate with your ecosystem of workloads. Ensure that the provider has also established mechanisms for planned and unplanned events that could potentially cause service disruptions.

Being aware of their processes and documentation will give you a full understanding of their limitations, and how to work with such limitations as an end user should be part of your planning as you migrate.

Service dependencies and partnerships

Ask for any **service dependencies** or **underlying partnerships** with other vendors to provide specific cloud services. Sometimes, SaaS providers build their services on other IaaS platforms, so you must get a full understanding of how their services are being delivered.

When running mission-critical business processes or having data that needs adherence to privacy regulations, you should look for providers disclosing these relationships across all parts of services that are being offered to get an explicit clarification and implications of SLAs.

Partnerships or vendor relationships are quite common among SPs and hence important to understand. Assess accreditation levels and understand if they support multi-vendor environments, and ask for good examples or use cases. It is vital to think about how these will fit into a larger ecosystem that can complement your business requirements.

Overall business health

We discussed earlier the importance of evaluating the technical and operational capabilities of a potential cloud vendor, and another aspect in conjunction with this is considering the financial health of your vendor. A vendor's **financial health**, in addition to contract assurances and any past legal challenges, is something that needs to be understood to get a sense of their market status.

Consider how the vendor is doing in their business for a long-term partnership. Evaluate the provider's **track record of stability** and their financial position to gain an idea of whether they could operate successfully.

Simultaneously, place an emphasis on doing your own research, such as reading up on any **case studies** of enterprises in similar business segments, as well as interviewing peers who are using the same vendor to understand their experiences.

Checklist for choosing a strong partner

Here is a comprehensive checklist of items that you can use as a strategy while evaluating and identifying the right cloud provider for you and your business:

1.   Validate compliance certifications and standards

2.   Evaluate case studies and testimonials of other enterprises

3. Interview peers for their experiences and lessons learned

4. Evaluate security controls

5. Understand the cloud provider's storage architectures

6. Evaluate the shared responsibility model

7. Understand capacity limits and compare with other providers

8. Understand the maturity levels of the provider's data center infrastructures

9. Evaluate the sustainability of the vendor's infrastructure

10. Assess the provider's technology capabilities

11. Ensure data retention for legal purposes

12. Ensure that your data resides in a physical location that has no regulatory requirements for access

13. Establish workable SLAs and contractual terms

14. Verify **user experience** (**UX**) performance

15. Verify network performance

16. Ensure an evolving service roadmap and technology innovation

17. Verify maturity levels of data classification and information security

18. Ensure key vendor relationships

19. Analyze service dependencies and codes of practice for implications of SLAs, accountability, and responsibility

20. Review contracts, renewal rates, and data protection policies

21. Review business terms, caveats for contract renewals, and notice periods

22. Compare costs across cloud vendors

23. Understand DR provisions, processes, and data preservation expectations

24. Evaluate technical support accessibility and availability

25. Evaluate a long-term roadmap and a clearly defined exit strategy

26. Compare business health and company profiles

Evaluation of potential vendor lock-ins

Make sure to include as many factors as possible in your assessment of potential providers, and recognize and compare the benefits and limitations. As leaders, you need to make sure that these areas are addressed as applicable, and do a thorough analysis before you make a long-term decision that impacts your business.

Unraveling the multi-cloud – benefits, challenges, and strategy

With multi-cloud deployments becoming increasingly popular, many companies are evaluating multi-cloud deployments. With the multi-cloud model of computing, an organization can leverage a combination of clouds. Many digital innovators and leaders want to achieve maximum flexibility and are investing in multi-cloud strategies, whether it comes to hiring staff or expanding efforts in automation software.

In this section, we will discuss if you should consider the multi-cloud model and explore what it looks like to manage infrastructure on more than one cloud provider.

Benefits

In this sub-section, we look at the benefits that organizations can reap if they choose to implement a multi-cloud option.

Improved DR and SLAs

Multi-cloud gives you the benefit of redundancy—that is, if one cloud service goes down, you can still have your mission-critical applications run on the other cloud provider and ensure that your data is not lost.

Many companies have requirements to keep their latencies low and even route traffic through the internet to different regions when a breach in their SLAs is noted. Enterprises that have stringent needs to keep improving their availability look for different cloud providers and a multi-cloud setup.

Innovate rapidly

Cloud providers are innovating at a rapid pace to meet customers' ever-evolving requirements and to stay as leading providers in the market. Developing new features and launching offerings help many businesses solve their problems efficiently. As a multi-cloud innovator, your business can leverage iterations of the latest cloud developments and improve your offerings.

Reduced lock-in risk

As we discussed earlier, enterprises can benefit from staying agile and ready to move easily between clouds as needed, with the standard approach of building applications and operating them on multi-cloud.

Specialized support and cherry-pick customizations

When it comes to cloud providers, there are striking differences in how their constructs and services are fundamentally built. The use cases that each of these services is solving could benefit from different requirements and even meet some of the custom needs of a business. It is easier to tailor features and functionalities when you have the flexibility to choose from multiple cloud services.

Diversification of use cases

Enterprises like to have choices when it comes to service offerings. Having the flexibility to prevail on any kind of platform will give users the freedom to run their applications on different technology platforms.

Next, let's take a look at some roadblocks or challenges to look for if you are considering multi-cloud for the first time.

Challenges

There are challenges surrounding multi-cloud infrastructure, and some of them can be daunting if you are entering the cloud world for the first time. Let's take a look at what these challenges are.

Management overhead

Moving from on-premises to the cloud itself is doubling your footprint in terms of understanding infrastructure constructs as well as supporting your applications on both platforms. Moving to multi-cloud means that you have tripled the work in terms of sheer management of resources, monitoring, provisioning, and maintaining this footprint.

Needs more budget

A multi-cloud setup depends on factors such as your architecture and the kinds of services that you want to integrate to be able to use each cloud provider. This can lead to building and maintaining a product that is compatible with more than one cloud provider, which can turn out to be more expensive than a single cloud provider. This requires a ton of work, and running in production on more than one cloud can double the efforts, as well as causing you to lose bulk discounts that are typically offered by CSPs when you split your workloads.

Learning curve

When you are running production systems on the cloud, you typically look to hire cloud architects, network engineers, automation engineers, security engineers, and developers.

Your teams will need to gain the required skillset to understand and operate on multiple clouds, and that can cause a lot of turmoil. It is expensive to hire talent that has such skillsets. Expecting an expert to run a multi-cloud solution can be a daunting task that many companies will not be able to sustain.

The following screenshot highlights the benefits and challenges of multi-cloud:

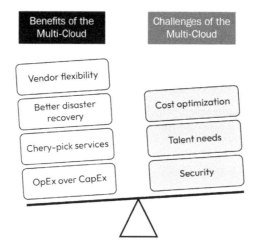

Figure 3.5 – Benefits and challenges of multi-cloud

Now that we have discussed the benefits and challenges, let's discuss some strategies for you to prepare your organization's IT landscape to be successful with multi-cloud implementation.

Multi-cloud strategy

We now know some of the benefits and challenges that come with multi-cloud infrastructure. If you decide to choose multi-cloud after considering all the preceding factors, make sure you incorporate the following recommendations to avoid problems in multi-cloud. These best practices can make your multi-cloud setup possible and easier to work with.

Modularizing your systems

Modularization is a technique for dividing your applications in a smart way where it is configurable in different combinations, and you can rebuild different systems with a given number of the same building blocks.

For example, choosing HashiCorp Terraform as your **infrastructure as code** (**IaC**) can help with provisioning on multi-cloud easily. This is because you know which infrastructure resources can be provisioned easily on both cloud providers, and you can recreate equivalent constructs using the same code snippets. You can reduce risks and increase the productivity of your teams given that you are eliminating the complexities associated with multi-cloud deployments.

Choosing a phased approach

When you start with one cloud, take careful consideration into account, and moving applications in an iterative way can make each step of yours a successive as well as an informed step. A good multi-cloud strategy is well thought out and mapped to your needs. Spend as much time as possible in the planning phase of your strategy and include security and data experts to ensure that you are identifying any risks or complexities. Having a well-crafted plan with the right teams included will ensure that your strategy is sustainable in the long term.

Data safety

Some infrastructure resources on the cloud that you choose may have requirements to ensure that your data is safely stored on multiple cloud vendor platforms. Knowing what is best for your data and looking at the granular level to ensure that your data is highly available and is stringently protected should be a top priority for you. Now that we discussed some key considerations that you must evaluate while choosing a cloud provider and have learned about multi-cloud benefits and challenges, let's look at how to prepare for cloud migration.

Aligning your IT landscape with the cloud – best practices

For many companies, an overarching question remains: *Where do we start when preparing for the cloud migration journey?*

There are a few guidelines when it comes to preparing your IT applications for the cloud. Regardless of the migration, these guidelines will help you make strides in your overall software development processes.

Your initial conversations will revolve around identifying your needs and establishing a business case, but more importantly, if your applications align with best practices such as those discussed next, you will be set up for a seamless migration.

Implementing SDLC best practices

Software Development Life Cycle (**SDLC**) is a model that helps you to perceive how software is built in a series of nine different functions, such as Planning, Requirement Analysis, Development, Testing, Documentation, Deployment, and Maintenance.

Not all the aspects of SDLC may be relevant for this section here but emphasizing a **development-operations** (**DevOps**) culture is highly recommended irrespective of the platform you are running your applications on. Practices such as IaC help you implement de facto solutions in development today.

IaC is a process of automatically provisioning and managing your infrastructure. This is done through writing code and allows the infrastructure to be easily integrated into code version-control mechanisms.

This helps in improving the efforts of your team through easy and efficient automation and ensuring results that are more reliable, error-free, and robust. Ultimately, you benefit from a reduction in **time to market** (**TTM**), and developers can advance their software to keep pace with changing needs.

When you inculcate best practices such as these in your organization, your teams are naturally getting ready for cloud infrastructure implementations, and you are closing the gap on the cultural shift as leaders.

Building portable applications

We talked about portability and its importance in the *Vendor lock-in criteria and exit provisions* section earlier. When it comes to portability, building loosely coupled systems becomes the core foundational aspect.

The following screenshot shows how a tightly coupled system can be different from a loosely coupled system. Binding resources to cater to specific use cases are very purpose-built and can bring in zero software extensibility and scalability. Opposite to this paradigm is loosely coupled systems that are detached from each other. All the systems can work independently in close concert with multiple components to deliver benefits in terms of extensibility and scalability:

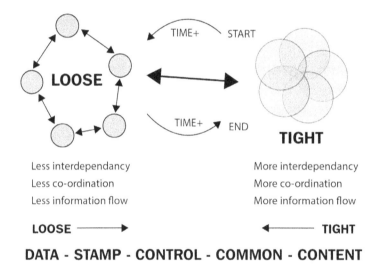

Figure 3.6 – Tightly coupled system versus a loosely coupled system

Let's discuss some key enablers for application portability, as follows:

- **Open APIs and standards** are the most common way to address the interoperability problem. When you design applications based on an open architecture using open APIs and standards, you achieve portability and promote interoperability. Look for vendors that have a clear mandate to specify and publish services through open APIs and standards. Some commonly used standards are **Open Cloud Computing Interface (OCCI)** and **Topology and Orchestration Specification for Cloud Applications (TOSCA)**, but there are also others.

- **Containerization** enables your applications to be run in a virtual environment by packaging your libraries and applications into a container. Containerization provides portability, flexibility, and scalability, making it possible for companies to run their applications in a vendor-agnostic environment. Containerization is the backbone for modernized applications on the cloud. We will discuss containerization in detail in the coming chapters.

Given that legacy applications form the crux of many businesses, it is key for organizations to consider the preceding discussed recommendations. We will dive deeper into containerization and building API-driven portable applications in the upcoming chapters. The takeaway is that IT teams should look for ways to connect to the digital world with minimal disruption.

Summary

In this chapter, we discussed how today's cloud ecosystem has a variety of factors to consider before planning and preparing for cloud migration. We learned about cloud migration insights that will help you plan better, move faster, and operate simpler. Multi-cloud and single cloud being the buzzwords, we broke these down and discovered their benefits, challenges, and how to develop a strategy to handle multi-cloud deployments.

We also looked at numerous factors that influence the decision to choose a solid CSP to help you innovate your business faster and deliver results that matter to your business. In the next chapter, we will dive deeper into cloud migration strategies so that you can design your migration journey with a better understanding of industry insights before you make the actual move.

Further reading

To learn more on the topics covered in this chapter, please visit the following links:

- https://www.iso.org/standard/67545.html
- https://standards.iso.org/ittf/PubliclyAvailableStandards/index.html
- https://www.snia.org/cdmi
- https://publications.opengroup.org/c163
- https://data.gov.ie/formats
- https://www.hashicorp.com/products/terraform/multi-cloud-compliance-and-management

4

Implementing Cloud Migration Strategies

As you prepare for transitioning to the cloud, it is important to gain a comprehensive view to achieve a successful enterprise cloud migration. Developing a strategy and having your teams use that as a guide will have an impact on your efforts. There is no one-size-fits-all solution when it comes to strategies, and it is important to identify what works best for you and adjust the strategy based on the results of the migration as you iterate.

In this chapter, we will discuss strategies that you can consider while migrating and how organizations should focus on key aspects that will facilitate the positioning and purpose of your migrations.

We will cover the following topics in this chapter:

- Introducing cloud migration strategies
- Diving into the 6 Rs of cloud migration
- Building a business case for cloud migration
- Mastering cloud migration using **Amazon Web Services** (**AWS**)
- Choosing a cloud migration strategy checklist

Introducing cloud migration strategies

In the prior chapters, we discussed how organizations can be blocked with their cloud migration efforts if they don't leverage a proper migration strategy. In this section, we will uncover what a cloud migration strategy is, what it entails, its benefits, and different types of strategies that you can incorporate into your cloud migration efforts.

What is a cloud migration strategy?

A cloud migration strategy is a **systematic** and **documented** plan that an organization can use to move their existing business elements, data, and application workloads from on-premises to a cloud platform. The plan typically addresses both *technical and cultural transformation* across the enterprise.

This detailed step-by-step migration plan is usually prepared by cloud architects who must have enterprise-wide visibility to understand the business well. This plan is then presented to stakeholders of the enterprise—such as leadership, technology, and security teams—to get buy-in and support throughout the process. Once top-down approval is achieved, the plan is used to train the team for the migration.

A solid strategy roadmap becomes the foundation for a successful migration that typically takes into account *business, architecture, timelines, application dependencies, and any contingencies*.

Purpose of a cloud migration strategy

Many organizations, large and small, are increasingly preparing to make the move, and using a cloud migration strategy will keep them on track and accelerate their digital transformation.

A well-crafted plan must maximize the value of organizations in the ways discussed next.

Identifying the scope, current state, and desired state

The most important element that the strategy should cover is the scope of the migration. It becomes easy to outline the scope if you understand the current state of your organization to inform and guide you to create a cohesive plan on the scope.

Which applications or workloads will be candidates for migration will become the crux of your scope. Outlining the business objectives and identifying workloads that cater to those objectives will become part of the scope. Make sure to include the desired state that talks about how the immediate plan is going to fold into the long term, including any criteria and **key performance indicators** (**KPIs**) that will be helpful to measure and report on successes.

Determining licensing, maintenance, and support contracts

While you formulate the scope of the migration, have the architects create an inventory of expiring licenses and contracts. This can be used to determine applications for which licenses need not be renewed if migrated and that can be prioritized in the plan. By doing so, the business can drive more value through termination, as well as save money on recurring legacy license costs.

Identifying dependencies

As you craft the scope of your strategy, you need to determine the current on-premises architecture and get a full understanding of the applications and their dependencies to reduce the possibility of any roadblocks.

Listing applications and their dependencies, underlying services, or inventory ahead of the migration will address any risks and help you create a well-navigated map of how to translate these dependencies—such as databases, external services, or third-party **application programming interfaces** (**APIs**)—to the cloud platform; for example, **HyperText Transfer Protocol** (**HTTP**) and HTTP/2 clients for Android and Java applications, Java libraries that can be used to convert Java objects into their **JavaScript Object Notation** (**JSON**) representation, and so on.

Determining on-premises hosting locations

One key lever that can be used to scope your strategy is whether your organizations are using multiple data centers to host your workloads and which specific data centers you are targeting to close first. Low-latency, local data processing, and data residency restrictions should also be taken into consideration while making this decision. Many cloud vendors today provide regions and **Availability Zones** (**AZs**) that are designed to eliminate any latency limitations, so you should rest assured that targeting impending data center lease terminations and prioritizing accordingly will be a pivotal moment for your businesses to retain the most value from the migration.

Prioritizing services

Prioritizing workloads and chalking out the benefits of migrating them is an exercise that you want to mandate for formulating a successful migration strategy. Target low-risk applications to migrate first versus workloads that may require regulatory or compliance requirements. This helps in not only bringing the most value from the migration but also accelerating your migration. While iterating through your migration process, don't forget to document common design patterns in order to create a repeatable approach.

Involving key stakeholders

As you define the roles and responsibilities of the various team during the migration, it is crucial to involve the stakeholders. The stakeholders can be from cross-functional teams within the organization, such as security, application development, finance, or infrastructure departments. The stakeholders will help you strengthen your plan by providing feedback and support during the implementation of the plan. Your migration project will typically be funded by an executive sponsor, and these stakeholders will extend support to make the migration a successful effort.

Identifying central governance

As you deploy and run workloads on a cloud platform, many organizations have found immense value in implementing a *cloud architecture team* that acts as a **cloud center of excellence** (CCOE) to support the organization's migration in an effective manner. The early establishment of such a governance body will help the organization in defining consistent policies across various different teams.

Identifying an exit strategy

Having a plan to move services and resources out of a cloud platform requires as much planning as the initial cloud migration. Doing this analysis upfront gives a detailed understanding before any services are provisioned on the cloud. Make sure to include an analysis of a different cloud vendor in your strategy, along with what it takes to come out of a specific cloud platform after the migration. Many organizations seek options such as a hybrid cloud to avoid *lock-ins* on one cloud provider.

Diving into the 6 Rs of cloud migration

When you start looking at migrating your existing applications to the cloud, there are six broad categories of migration. In this section, we will dive deep into all of these options and explain how to decide which migration strategy to choose for your applications. The following diagram refers to the six ways on AWS cloud.

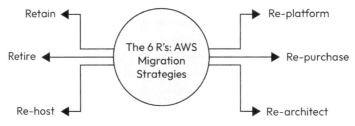

Figure 4.1 – 6 Rs of cloud migration

Rehost (lift-and-shift)

This cloud migration approach, also known as *lift-and-shift*, is the strategy that most companies gravitate toward given its relatively low effort. With this approach, the application and its dependent resources are simply moved to a cloud platform as-is.

This means that your core framework of infrastructure components, your application's dependencies, and the end-user experience remain the same. You will be able to meet scaling needs better as you begin to leverage the scalability and elasticity aspects of the cloud platform.

As you rehost your applications on the cloud platform, you can benefit from **operational expenditure (OpEx)** savings in the cloud by iterating through your workloads to consume cloud offerings in domains such as network, compute, or storage. Your long-term goal to operate in the cloud should be to optimize as much as possible and ultimately be future-proofed to see a continuous reduction of OpEx.

The following diagram shows how rehosting works:

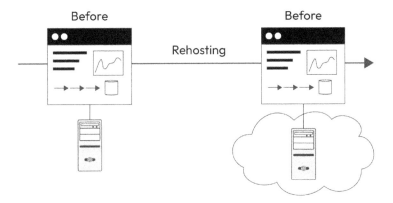

Figure 4.2 – Rehost

Let's take a look at a scenario for leveraging rehosting.

> **Scenario**
>
> A large media company is running its application on a **virtual machine (VM)** in a data center. Provisioning the application manually was labor-intensive and required startup scripts. With this migration, the application is moved to a cloud provider's compute instance. The company was able to gain complete control and fully automate its compute, thereby reducing the time required to spin up new instances in minutes and scale capacity both up and down.

Use cases

Let's take a look at some use cases for a rehosting strategy.

Large-scale enterprises

This is a low-risk effort compared to other migration strategies and is favorable for organizations that want to quickly move their legacy applications and leverage the cloud's scaling benefits.

Need for quick migrations

When speed matters for your organization or there is an immediate event such as termination of a data center lease, rehosting can be the fastest option, especially when you are leveraging automated tools such as CloudEndure, AWS VM Import/Export, or Amazon Lightsail. However, manually executing these migrations equips your teams with the required skills to carry over migrations with ease.

New to cloud migrations?

If you are new to the cloud, rehosting enables you to experience and utilize the advantages of the cloud without the need to redesign your workloads. This gives you the comfort of minimal disruptions on all fronts and gets the best **return on investment** (**ROI**) without making extensive changes.

Conclusion

Rehosting is often misunderstood as a process that is a *NO-FAIL*, which is not the case as rehosting requires proper planning and understanding in terms of the required process changes, configurations, and differences in how workloads operate on a non-cloud platform versus a cloud platform. Identifying any security-related issues by conducting penetration testing for your workloads on the cloud (as well as post-rehosting activities) should be taken care of to ensure that you do not encounter any problems once you migrate.

Replatform (lift-and-shape or lift-tinker-and-shift)

Replatforming is one level up from rehosting. It is important to consider a few cloud optimizations if your organization is keen to see tangible cloud benefits. This does not necessarily mean that you have to completely redesign your application end to end.

With replatforming, you selectively identify a set of changes in the ways your applications interact and take benefit of the managed platform services that **cloud service providers** (**CSPs**) offer. The idea is to keep the *risk factor minimal* but look at minor changes that will bring you improvements in the overall stack of your application.

By swapping some components within your stack, you get to minimize the footprint of processes that you get to manage, thereby reducing OpEx and benefiting from *cost and performance improvements* from the CSP offerings without spending time rewriting the application.

The following diagram shows how replatforming works:

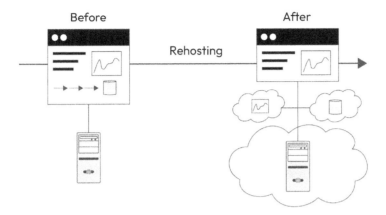

Figure 4.3 – Replatform

Let's take a look at a scenario for using a replatforming strategy.

Scenario

A marketing firm is running an application connected to a relational database on instances hosted in a data center. With replatforming, they move to a CSP's compute instance and swap the database component with the CSP's managed database service. By using the CSP's managed service offerings, the firm is able to save on OpEx required to monitor and scale their databases, thereby enabling the firm's engineering teams to focus more on application development.

Use cases

Let's take a look at some use cases for a replatforming strategy.

Short migration timelines

Making **cloud-friendly choices** enables teams to reduce the time it takes to build these components instead of bare-metal instances that could be manual configurations and executing fragile scripts that are often error-prone.

Benefiting from cloud-native

By enabling some common components of your application with cloud-native services, you can drive the business to achieve cost and performance improvements and reliability. These benefits can be attained with minimum risk and less time and make it a choice for many business leaders.

Conclusion

With replatforming, your applications that are consuming cloud services result in less management overhead, higher availability, and elasticity. Understanding the equivalent cloud services while replacing application components is a crucial factor to ensure that replatforming is done right. While reshaping may increase the risk of causing errors, well-known replacements will lead to a successful replatform and thereby produce a better solution overall.

When you choose replatforming, ensure that you have a team of skilled staff, including architects, engineers, and consultants who are comfortable with cloud-based technologies. It is recommended that organizations invest in continuous workforce development as they consume more cloud offerings.

Repurchase (drop-and-shop)

This strategy, often called **drop-and-shop**, is the quickest way to run applications in the cloud, whereby you eliminate the migration effort of the applications and hence the migration risk. You simply drop the existing on-premises workloads and their components to start with a new *user/license agreement* and set up your applications directly on the cloud.

You still have to take care of migrating data; depending on the requirements, this could be accomplished by *third-party tools* to quickly do so automatically. Make sure to add this to your planning step if you choose this approach in order to assess the required effort.

Understanding new services requires your staff to be well equipped with the available controls and customizations. The key is to determine how much of a tailored solution you would need to balance the requirements with readily available software that offers better value with higher efficiency and lower maintenance costs.

The following diagram shows how repurchasing works:

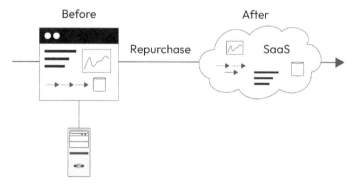

Figure 4.4 – Repurchase

Let's take a look at a scenario for using this strategy.

> **Scenario**
>
> An online real-estate investment company offers its employees to connect to the company's network through a self-built **virtual private network** (**VPN**) server. The company was using standard functions but paid more to maintain the routers. They swapped to a fully managed elastic VPN service offered by a CSP to eliminate the need to install and manage hardware or software.

Use cases

Let's take a look at some use cases that would align with a repurchase strategy.

Third-party dependencies with no vendor support

With the rapid increase in open source, many third-party offerings do not come with any vendor support. By letting go of past lock-ins and replacing them with modern **software-as-a-service** (**SaaS**) offerings that are developed by specialists that you do not need to hire on your team, the migration components are further reduced, and it makes it cheaper and less risky for your migration program.

Systems relying on standard functions

Your legacy or enterprise software that did not have any updates in the functionality in recent times is a good candidate to be swapped with cloud offerings that offer a standard functionality stack that is frictionless and easy to operate with continued operational support.

Conclusion

As you see a rise in the trend of moving to a consumption-based purchasing model, many organizations are leaning toward **no commitment pay-as-you-go** models. With this approach, you don't have to deal with **capital expenditure** (**CapEx**) but rotate your software stacks more often to remain innovative and suitable for your businesses.

Refactor/Rearchitect

Refactoring or rearchitecting involves redesigning your application using cloud-native offerings. This is an approach that entails rewriting from scratch to realize the full potential of cloud-native offerings for your workloads. As you select applications for potential refactoring, make sure to look at detailed activities and outcomes to understand the overall impact.

Technologies such as **serverless and containers** are the most common computing upgrades that businesses tend to let go of in their traditional applications. These traditional applications often come with heavy overload and less flexibility and give a good reason to be deconstructed and decoupled for **fine-grained** cloud consumption.

Chief technology officers (**CTOs**) will typically rely on architects to make a **strong business case** for a refactor because this approach can get expensive, requiring cloud-skilled resources when compared to the other approaches. Additionally, there is an effort from the development team to support the current application as well as refactor code that will be running on the cloud. Depending upon the complexity, it is recommended to have a project delivery team. The team helps in managing the project with an Agile methodology to achieve a successful business outcome.

The following diagram depicts how refactoring works:

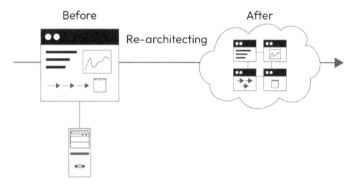

Figure 4.5 – Refactor or Rearchitect

Let's take a look at a scenario where a refactor strategy works.

> **Scenario**
>
> An e-commerce company wants to revamp its shopper experience for speed and scale. In order to ensure they can support an ever-expanding business, they would like to refactor their monolithic application and migrate to microservices, as well as leverage a cloud data storage and **machine learning** (**ML**) engine to recognize shopping patterns. In addition, the company moved to Docker for stateless configuration management and used a cloud container orchestration service to offload cluster management.

Use cases

Let's take a look at some use cases that would align with a refactor strategy.

All-in on cloud

Organizations that have chosen a specific cloud vendor that leans toward moving their business elements, data, and applications choose this migration strategy. **Rebuilding** legacy systems by replacing them with their equivalent cloud services allows companies to achieve a long-term ROI where the emphasis is on scalability, speed, and performance.

Re-imagining business continuity

The ability of your application components to not be severely impacted by an unplanned incident is critical for your business. The benefits of using all cloud-native services for your applications through rearchitecting help businesses achieve operational resiliency. With the right levers—such as mechanisms for backup and restore on the cloud—**disaster recovery** (**DR**) scenarios are well executed upon an incident, thereby ensuring an effective **business continuity** (**BC**) plan.

Long-term cost benefits

Businesses that are looking for resource consumption on-demand with a breakdown per transaction can benefit from rebuilding their applications on the cloud to take long-lasting cost benefits of fully cloud-native systems.

Improved resiliency

If your current application's components are tightly coupled, it can get complex to troubleshoot your systems in the event of failure. Rearchitecting can be an ideal choice in such scenarios, whereby you decouple the components within a system and plug in highly available managed services that incorporate the resiliency posture inherited from the cloud ecosystem.

Realizing cloud innovation

In addition to the benefits of cloud resources, organizations can take advantage of the technology and innovation when compared to traditional life cycle management. Inheriting cloud advancements will help your businesses to be positioned well for continuous innovation and rapid improvements by choosing this strategy.

Conclusion

A rebuilding strategy is not for beginners in the cloud. This approach can have the highest risk of going wrong, so ensuring that your discovery assessment is solid (depending on the complexity) is key. To get the most benefit from this strategy, make sure you invest in the right skilled team and spend the required time while transitioning your workloads from on-premises to cloud-native.

Retain (do nothing)

As you assess your **information technology** (**IT**) portfolio, there will be applications that you mark as **not ready to migrate** at that point in time and continue running them on-premises. **Retaining** such workloads gives a sense of the priorities for migration. Understanding applications that are suitable to continue on-premises will help you prepare well for compliance or regulatory controls that you need to continue with.

The reasons for such decisions can vary, from the application not being ready to adopt cloud offerings in its current state, to having recently gone through upgrades of licensing that may otherwise turn out to be expensive to make the shift to the cloud. An organization can make an informed decision by taking into consideration what suits your business best overall.

The following diagram shows how a retain strategy works:

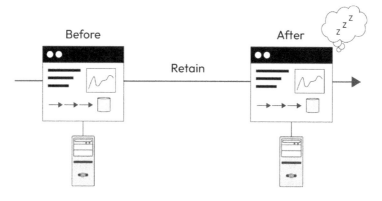

Figure 4.6 – Retain

Let's take a look at a scenario for a retain strategy.

> **Scenario**
>
> A banking firm wants to ensure continuity to its business while doing large-scale migrations that take years. In such cases, it retains some of its mission-critical workloads that carry confidential data on-premises to ensure a balance between on-premises applications and their ongoing migrations to the cloud.

Use cases

Let's take a look at some use cases that would align with a retain strategy.

On-premises investments

Companies that recently invested heavily in procuring hardware/software licenses to power their legacy applications typically choose to retain their applications until **end of life** (**EoL**) of those specific licenses.

Incompatible with cloud setup

If you are running old legacy systems that have dependencies on custom code that is not well suited to the cloud or there are no equivalent cloud offerings, you choose to retain such workloads instead of reinventing the wheel.

Regulatory requirements

If you are running applications that have strict compliance regulations and require your data to be kept on-premises, you will have to continue with a retain strategy.

Conclusion

Although cloud migration can be an answer to many of your IT operational challenges, it is always important to identify applications that do not benefit from such migrations in their current state. Some of the factors that we discussed can help you include this in your planning and put the migration of such applications on the back burner.

Retire (drop)

With this strategy, you mark the applications as **retire** if they are no longer adding value to your business. Decommissioning such applications allows you to cut down on costs to maintain them, and your staff can focus on migrating critical applications or services, reducing the overall effort and saving on total time to migrate.

The following diagram shows how a retire strategy works:

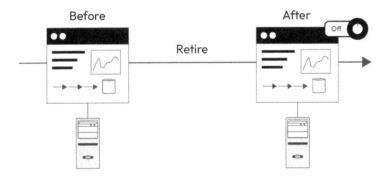

Figure 4.7 – Retire

Let's look at a scenario for a retire strategy.

Scenario

A manufacturing firm accumulated hundreds of applications through its recent acquisitions. While migrating to the cloud, the firm identified all the applications as having duplicate capabilities and retired them. This helped the overall business to reduce migration efforts and attack surfaces with better security and an improved end-user experience.

Use cases

Let's take a look at some use cases that would align with a retire strategy.

Outdated applications

Many enterprises over time don't realize that they have many applications that are outdated and vulnerable to security attacks. Such applications should be retired from time to time to ensure stronger security and efficiency.

No production value or redundant functionality

When there is no business outcome from applications, it is important to retire them and save costs. Identifying duplicate functionality across an enterprise's IT portfolio also helps eliminate unwanted costs that could be otherwise invested in the company's innovation.

Conclusion

The process of discovering your applications and interviewing each functional key stakeholder that owns that workload always gives businesses an opportunity to save and redirect their staff to do things that are more user-driven and reduce the operational overhead.

Building a business case for cloud migration

We have now established that every organization needs to carefully consider cloud migration. For this reason, a business case can represent the cloud migration opportunity from the lens of technical and financial aspects of your environment. If these considerations align with the business outcomes, this helps you get support from all the areas of business, thereby helping to accelerate the purpose of cloud migration.

The business case for cloud migration needs to be compelling to shake up any organization that is operating on a traditional on-premises model with a legacy mindset. Building a strong business case is not only important to persuade and get buy-in from key stakeholders but also helps you think about factors such as timelines, costs, and challenges it will solve early in your cloud journey.

Understanding how cloud migration fits in with your company's goals and vision will enable you to position your business case in the right manner.

Strategizing and crafting an efficient and effective business plan is crucial for bringing attention to the opportunity. Let's take a look at how to build a business case for the cloud.

Executive summary

An executive summary is a brief introduction of your plan, stating the business challenges that your organization is currently facing and how cloud migration will help address these challenges. Clearly state the purpose of migrating to the cloud, along with additional data points such as company description and current market analysis (if applicable). This section should be concise, compelling, and attention-grabbing to attract readers.

Defining a problem statement

Understanding why you are presenting the business plan, your company's current challenges, and the goals of the migration will be the best way to present the problem statement. It is important to personalize the problem statement with as much relevant data as possible so that you can highlight focus areas. These focus areas will then become core aspects that you would want to achieve through cloud migration. Cost savings and business innovation are some of the topics that every organization can focus on.

Gathering business data

CapEx and OpEx are terms that you would want to explain when it comes to including information on your current finances. It is always important to call out financial elements when you build your problem statement. Run your analysis before you present this data to ensure that your projections forecast cost savings.

The core of your IT technology footprint and long-term growth for your applications, along with the demand from your customers and aspects such as development and DR, will be factors for calculating your costs.

Security, resiliency, monitoring, performance, and strategic values are all additional common trigger points or drivers that C-levels of every company would be interested in to learn if cloud migration will help with those areas.

The following table serves as a template that you can use to capture data for this area:

Trigger area	Business unit (BU)	Business owner
Domain category	Specific line of business	Sponsor of the business

Table 4.1 – Gathering business information matrix

Next, let's look into the specifics of outlining the objectives of a migration plan.

Outlining your main objectives

Once you gather the pain points specific to your current environment that will benefit from cloud migration, you will need to specify clearly how your organization will look once the migration is completed. Mapping objectives and the target area they solve is a good exercise to complete for you to effectively write this section.

The following table is used to list the expected business outcomes:

Objective #	Outcome	Capabilities
xxx	List the required outcomes	List the necessary capabilities

Table 4.2 – Objectives matrix

Next, let's look into summarizing the proposal of a migration plan.

Summarizing the proposal

This is the section where your implementation and inner workings need to be detailed. Also, adding to the core of your strategy is how your partners' services will play a key role.

Before you summarize a list of applications, answer the following questions with an affirmation for you to be able to determine the best candidates for your proposal:

1. Is the application a market differentiator?
2. Is there an investment in this application to improve the experience?
3. Does the data for this workload support migration?
4. Will the business objective for this application change during the course of migration?
5. Is the source code supporting this application stable to support the growing business needs of your business?
6. Is innovation limited to the application's current architecture or operational overhead?

Outlining answers to the preceding questions will help you summarize a proposal for each of the applications of your IT environment and strategize the migration accordingly.

Highlighting limitations and risks

The potential risks associated with cloud migration and a detailed plan to mitigate them should be the core of this section.

As you understand more about your infrastructure, you will identify these risk areas easily. If not, it is worth spending time going through aspects that you may have ruled out along the way in a detailed manner.

Aspects such as data security, privacy, service disruptions, and cloud partner risks are some things that you should plan to take on in this section. Some other examples are detailed next.

Poor skillset rates

When it comes to cloud migration, one common risk is you will not be able to find enough resources to train employees ahead of the migration. IT team members will feel insecure about their positions and jobs as they realize that a different skillset is required to implement and support cloud migrations.

You can mitigate this by nominating a few experts who can act as team leads. Engage them and get their buy-in and full visibility to drive the team training so that they can become skillful and control migrated applications.

The following diagram shows a mapping of existing skills to cloud skills for your organization to get started with the skill-readiness journey to mitigate this risk:

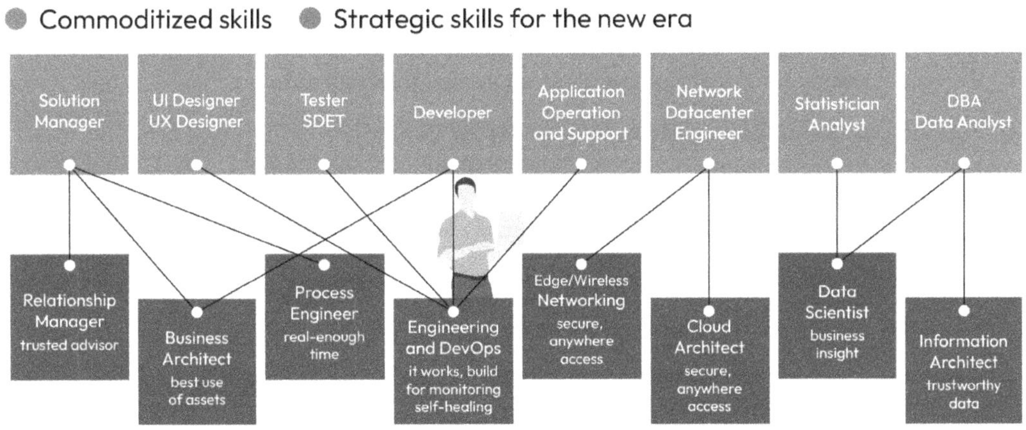

Figure 4.8 – Mapping of skills to IT roles in a cloud ecosystem

Service disruptions

It is important that you choose a cloud provider that can guarantee a certain **service-level agreement (SLA)** and uptime for your solutions to be available 24/7. Your applications should be well designed, and your cloud partner must be great with reliability so that any service disruptions do not negatively impact your business. We will cover this in high detail while discussing AWS Well-Architected Framework design principles and best practices to achieve this.

Outlining a migration plan

This section should focus on how objectives are converted into actionable plans. The technical efforts put in by your assigned cloud migration teams should align with your business objectives.

It is always important to break down your migration into phases so that you can learn as you iterate and decide the best strategy to achieve a successful transition to the cloud.

Your plan should list out core components so that you are calling out the migration phases and the targets of each phase in this section clearly.

The following table serves as a template for you to use while outlining a migration plan:

Migration phase#	Application name	Migration priority (high, mid, low)	Proposed outcomes
List the iteration number	List the name of the application	List the priority	List outcome

Table 4.3 – Migration planning matrix

To conclude, a business case should cover the following aspects:

- Current operational costs of running applications and infrastructure
- Projected costs of each key stage of your cloud migration
- Estimated cost savings from moving to the cloud
- Advantages in areas beyond costs, such as security and performance
- Advantages in the overall experience of your existing customers
- Advantages in speed to provision new services
- Advantages in satisfying on-demand surge capacities through elasticity and scalability
- Greater focus on newer technologies and innovation
- Advantages in financial performance and a basis to attract new business and achieve a competitive advantage

Now that we have discussed what it takes to formulate a good master migration plan, let's take a look at which tools and services are available to execute a successful migration plan using AWS services.

Mastering cloud migration using AWS

AWS offers a wide range of migration products, and this section will focus on how each of the AWS services shown in the following screenshot can help organizations of all sizes to migrate their workloads to the cloud:

Type of Migration	Tools
Discovery and Migration Tracking	AWS Application Migration Service
	AWS Migration Hub
	AWS Application Discovery Service
	TSO Logic
Server and Database Migration	AWS Service Migration Service
	AWS Database Migration Service
	VMware Cloud on AWS
	CloudEndure Migration
Database Migration	Amazon S3 Transfer Acceleration
	AWS Snowball
	AWS Snowmobile
	AWS Direct Connect
	Amazon Kinesis Firehose
	DataSync

Figure 4.9 – AWS services for cloud migration

AWS Migration Hub

AWS Migration Hub provides you the tools to store, plan, execute, and track your IT asset inventory data migration to AWS. With this service, you can gather information such as actual usage, application components, and infrastructure dependencies of workloads. AWS Migration Hub generates optimal migration strategy and recommendations for your business case to migrate and modernize at scale. Its simple and intuitive migration dashboards show metrics and the latest statuses of your migrations so that you can better understand their progress as well as identify any issues early on.

AWS Application Discovery Service

AWS Application Discovery Service lets you capture data about your IT environments specific to its configurations, usage, and behavior. Your on-premises infrastructure information, such as server hostnames, **Internet Protocol** (**IP**) addresses, **media access control** (**MAC**) addresses, resource allocation, and utilization data such as **central processing unit** (**CPU**), network, and memory, are discovered through the service's agent or agentless-based discovery process. This data is then used to understand dependencies across the servers and measure the servers' performance. Enterprises can get a snapshot of the current state of their servers and use it to generate **total cost of ownership**

(**TCO**) analysis. Application Discovery Service uses AWS Migration Hub as a single repository to store your discovery and planning data.

AWS Application Migration Service

AWS Application Migration Service is a migration service for lift-and-shift migrations to AWS where you can automatically convert your source infrastructure to cloud infrastructure on AWS. You can perform non-disruptive tests before the cutover of your application. You can use AWS Migration Hub to integrate and monitor your migration status.

AWS Database Migration Service

AWS **Database Migration Service** (**DMS**) enables organizations to migrate their commercial or open source databases to AWS. DMS automates your database migration planning so that you can migrate to the cloud at scale with minimal effort. DMS automatically assesses your on-premises database or analytics servers and identifies potential migration paths with no downtime. You can build a migration path and save on costs associated with the manual efforts of planning and migrating your workloads with this service. Databases such as Oracle, Microsoft SQL Server, MySQL, MongoDB, and many more are supported for migration.

Amazon VM Import/Export

VM Import/Export offers ways to easily import VM images to Amazon **Elastic Compute Cloud** (**EC2**) instances and then export them back to your on-premises environments. You can easily migrate your existing applications to Amazon EC2 by not altering any of your existing software or configurations. You can easily back up your data and replicate your applications to take advantage of autoscaling and elastic load balancing and all the other things that AWS has to offer to support your migrated workloads.

AWS Marketplace

AWS Marketplace is a curated digital catalog of solutions to migrate to the cloud at every stage, including any cloud migration strategy. You will find tools for migration discovery, planning, data migration, application profiling, cloud cost management, and much more. You can browse through tens of thousands of products that are published by several independent software vendors to find, test, buy, and deploy the tools you want. The controls that you can get from the marketplace are spread across many domain categories such as security, data, and analytics, across industries such as healthcare, public services, financial services, and so on.

For more information on these services, please refer to the service links that are provided in the *Further reading* section.

Choosing a cloud migration strategy checklist

Here is a checklist that you can use to drive a good cloud migration strategy plan. You can add any organization-specific items on top of this checklist, but the following set of items will help you get a good kick start:

- Do you have the scope of the strategy?
- Do you have alignment with the larger strategic initiative of your organization?
- Are the business objectives driving your migration clear?
- Do you have high-level outcomes mapped to the business objectives?
- Do you have the right migration team?
- Did you pick the cloud provider?
- Is the current state of your organization explained clearly?
- Is the desired state for cloud migration and adoption along with the KPIs to measure and report on the success clear?
- Do you have a high-level path to cloud migration?
- Are risks called out clearly?
- Are there any vendor lock-ins?
- Did you assess the cloud migration costs?
- Are stakeholders of the cloud migration identified?
- Do you have an end-to-end look at your cloud migration?
- Do you have an exit strategy?

Further reading

- Forrester report on cloud migration: `https://www.redhat.com/cms/managed-files/cm-forrester-tlp-cloud-migration-actively-embraced-analyst-paper-f9811-201711-en.pdf`
- AWS Marketplace: `https://aws.amazon.com/mp/marketplace-service/overview/`
- AWS Migration Hub: `https://aws.amazon.com/migration-hub/?p=ft&c=mg&z=3`
- AWS Application Discovery: `https://aws.amazon.com/application-discovery/?p=ft&c=mg&z=3`

- AWS Application Migration Service: `https://aws.amazon.com/application-migration-service/`

- CloudEndure: `https://console.cloudendure.com/#/register/register`

- CloudEndure documentation: `https://docs.cloudendure.com/CloudEndure%20Documentation.htm`

- Amazon Database Migration Service: `https://aws.amazon.com/dms/?p=ft&c=mg&z=3`

- Amazon VM Import/Export: `https://aws.amazon.com/ec2/vm-import/`

- Amazon Lightsail: `https://aws.amazon.com/lightsail/`

- User guide on VM Import/Export: `https://docs.aws.amazon.com/vm-import/latest/userguide/vmimport-image-import.html`

Part 2:
Cloud Modernization –
Application, Data,
Analytics, and IT

This part creates a baseline on cloud modernization for businesses to take their cloud journey a step ahead and drive growth. We will deep-dive into why modernization matters for any businesses that has invested heavily in building its legacy systems. We will cover application modernization patterns and the tools available in **Amazon Web Services** (**AWS**) for building modern applications on the cloud. You will learn about application modernization strategies, design patterns to decompose monolithic applications into microservices, and the best practices to apply while building a plan for your modernization efforts. We will look into application integration patterns for microservices and review asynchronous messaging for building loosely coupled systems at scale. We will introduce data infrastructure modernization and discuss its importance in today's world, where data is everything when it comes to making business and technology decisions for data-driven companies. Data modernization techniques and patterns will be discussed along with the services available on AWS to build a highly scalable and secure data platform.

This part comprises the following chapters:

- *Chapter 5, Modernization in the Cloud*
- *Chapter 6, Application Modernization Approaches*
- *Chapter 7, Application Modernization – Compute*
- *Chapter 8, Implementing Compute and Integration on the Cloud Using AWS*
- *Chapter 9, Modernizing Data and Analytics on AWS*

5
Modernization in the Cloud

This is an exciting chapter where we will enter a phase that many enterprises envision – to modernize and innovate for their customers. Cloud computing has gained a lot of buzz in the industry, and by now, you should be familiar with the various features and offerings that a company can leverage and get a **return on investment (ROI)** from. We have also looked at different service providers for you to choose from, including **platform as a service (PaaS)**, **software as a service (SaaS)**, and **infrastructure as a service (IaaS)**.

Transforming in the cloud begins here – not only will you gain a deeper understanding of the cloud but you will also get to decide what your journey in the cloud will look like. Cloud modernization can empower you toward true transformation at an enterprise scale to deliver cutting-edge technology and accelerate innovation.

In this chapter, we're going to cover the following main topics:

- Introducing cloud modernization
- Uncovering the stages of modernization
- Understanding migration versus modernization
- Exploring the benefits of modernization
- Getting started with modernization on AWS

Introducing cloud modernization

The cloud is a key driver for innovation and is the foundation for many enterprises. As companies are evolving in their cloud migration journey, many are actively exploring new ways to reap business benefits by reducing the cost of operations and achieving business agility. Technical leaders are pursuing full-scale modernization to ramp up their organizations' speed and reduce technical debt.

Modernization is a continuous optimization approach for companies to apply and incorporate and transition their applications, data, infrastructure, and services to an enterprise-ready cloud.

Since the majority of enterprises are embracing the new cloud paradigm across their software design, development practices, and deployment methods, taking advantage of cloud modernization enables them to maximize the impact and improve the customer experience.

Modernization is not always easy, and some of the biggest challenges include the most basic questions:

- **Complexity**: Where do we start with modernizing and what do we modernize?

- **Cost**: How do we justify the time and money it takes for modernization?

- **Skill sets**: Do we have the right people with the right skills to deliver cloud modernization?

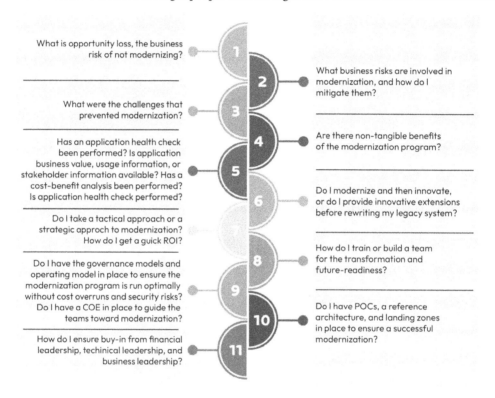

Figure 5.1 – Top questions to ask

To answer these questions and overcome any challenges, you need to connect the cloud to business outcomes. Addressing these questions early in your modernization journey will enable you to make cloud modernization a *business imperative*, rather than just a nice-to-have:

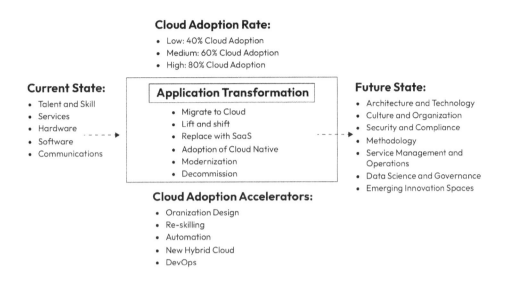

Figure 5.2 – The business's key performance indicators (KPIs)

Let's take an example of a media and entertainment company, ABC, a champion online streaming platform that airs ad-free content for users to watch anytime and anywhere over the internet. Amid the COVID-19 pandemic, the amount of consumption increased at a rapid pace with the free time that consumers had during the lockdown across the world. To meet this growing demand, ABC's cloud team redesigned its streaming platform on the cloud.

This enabled them to store their videos in different locations throughout the world and deploy a **content delivery network** (**CDN**) of distributed servers, which delivers content to a user based on their geographic location for faster load times and reduced latencies. In addition, ABC leveraged a fully managed load balancer service to route traffic across instances on the cloud.

With these features of the cloud's managed service offerings, ABC was able to benefit from less expensive, better-quality, and more scalable streaming and did not have to worry about the heavy lifting of the operational burden.

The secret to ABC's success was its understanding of how to leverage and maximize business results via the cloud; this often stretches way beyond IT activities. Many enterprises expect positive outcomes from technologies but neglect to foster the business case, skill sets, or culture to achieve the desired results. Let's take a look at some of the aspects that fuse all these dimensions to see the magic unfold.

The road to cloud modernization

The process of modernization needs thorough planning for a sustainable outcome in terms of business continuity, automation, security, and scalability for your IT footprint. Embracing cloud modernization requires substantial investments upfront and active commitment to a top-down approach for a long-term all-in transformation. Before we learn how to get started with cloud modernization, let's see whether it is the right time for your enterprise to consider cloud modernization.

Is it the right time to modernize?

With many enterprises striving to become first-to-the-market producers and meet the demands of digital mass penetration, it has become imperative that they have flexible and scalable IT setups to innovate at pace. In addition to the factors that we discussed in the previous section, make sure you perform an up-close evaluation of factors such as those shown in the following figure. It is important to note that end-to-end changes are dependent on variable factors such as the size of the organization and its vision and technical expertise:

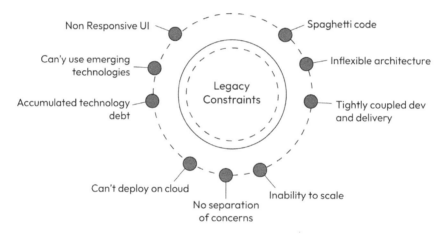

Figure 5.3 – Evaluators to modernize

How you transform your digital footprint can become the core of your future legacy systems. This makes many leaders seriously consider their technology modernization strategy.

To start with, you may want to address the following key elements to successfully begin your cloud modernization journey:

- **Look beyond lift and shift**: Lift and shift is often the go-to option for many enterprises as it is less resource-intensive and you can move your workloads to the cloud without tinkering with any code. Most of the time, this does not allow you to align with the security, capacity, performance, or resiliency requirements of your enterprise. The infrastructure and application layers of your existing applications need to be tweaked so that you can create a clear alignment and achieve legacy application modernization.

- **Build the business case**: Showing your company's internal stakeholders and leaders how each of their teams and the overall business will benefit is the core part of building the business case. Once you can garner their support, the cloud will become their priority. You can convince your stakeholders by taking a *business-focused approach* and targeting the most critical business applications in your business case.

  In the previous chapter, we discussed the areas for building a good business case, such as the problem statement, technology stack, project description, and benefit analysis. Additionally, as a leader, you may want to build a strategy for modernization by structuring your business case to reflect a continuous modernization approach to ensure all your applications are keeping pace with modern technology and aligning with your business purpose.

- **Hire/develop the right talent**: Having the right skills to design, develop, and deploy applications on the cloud securely and quickly is the key. To do this, enterprises will need to invest in hiring or training cloud experts and use them to upskill the existing cloud teams. Cloud development can be achieved via training and hands-on digital innovation workshops.

  Leadership needs to commit to investment because executing cloud transformation requires multiple things to be changed at the same time. Ensure you have a core team of cloud subject-matter experts that your teams can go to for auditing the cloud environment. This team can be expanded as the migration footprint grows and can retain invaluable lessons from transformation. With strong leadership, this hurdle can be significantly managed and become the catalyst for transformation.

- **Cultivate a cloud-first culture**: Cloud modernization is a journey on which organizations need to believe in a cloud culture – a culture of embracing innovation and transformation. Having a fail-fast mentality is the foundational step for achieving resilience and business growth through your applications. Real transformation happens when the people, processes, and technology are aligned. As a technology leader, you will need to constantly inspect and stay on top of questions such as the following to align with cloud modernization:

  I.   Do I need to refresh my workforce's skillset?

  II.   Do I need to re-engineer processes and workloads?

  III. How do I measure employee value in the modernization era?

  IV.   Do I need to restructure organizational processes?

The answers to these questions can impact your modernization pace and can require commitment, investment, and agreement from an organization's leaders.

It is also important for many enterprises to develop a cloud technology stack and a strategy to maintain it. Partnering with cloud providers is a typical approach where you start with a single cloud provider and adopt the required guiding methodologies to avoid being locked into one provider.

For enterprises who are lacking organizational buy-in or are reluctant to invest in a multi-year effort, strategize modernization in iterations that take short-to-medium timeframes. That way, your organization will still be able to build the foundational capabilities and can prepare for cloud modernization when your teams are comfortable and ready.

Pursuing modernization of segments of the end-to-end application life cycle such as automation or DevOps is also a common pattern before diving all-in with modernization. Other emerging trends are **site reliability engineering** (**SRE**) and **centers of excellence** (**CoEs**) for enablement and dynamic and self-sustaining independent teams.

SRE is about building highly resilient platforms and applications where services remain independent and running, even if a dependent service suddenly goes down.

- **Prioritize modernization efforts**: Modernizing everything at once may be a nirvana state but not realistic. It is important to begin the modernization efforts in phases, where you start with discovery and evaluate your existing IT portfolio and prioritize cloud modernization. As you dive deep into the workloads, think from both technical and business perspectives and classify your applications based on what their current state is and how the projected future state can bring in business value. Understand the interdependencies before classifying these applications.

  Historically, many enterprises encounter challenges with competing business demands. Consider modernizing your core business applications on the cloud to leverage the latest technologies such as purpose-built databases, containerization, microservices, serverless computing capabilities, and machine learning for effective modernization.

Diving deep into these key elements will help you remove any friction points and set the right expectations for your enterprise to be empowered and rethink its modernization approach across business and technology. Building top-down cloud modernization advocacy can generate momentum for the transformation and act as a catalyst for innovation. Now that we've discussed the key elements for moving toward modernization, let's take a look at the key steps that lay the path to modernization.

The five key steps

According to the *2021 Mainframe Modernization Business Barometer Report*, 77% of organizations have started but failed to complete at least one modernization program. One of the primary reported reasons was the disconnect between business and technical teams. The research found that **Chief Information Officers** (**CIOs**) and leaders are more interested in their organizational technology landscape, while enterprise architects are more focused on their specific teams.

As a technology leader, you need to collaborate and develop a cloud modernization strategy with your business to harness technological changes to reap successful modernization. When you are evaluating opportunities for cloud modernization, there are five steps: *Align, Design, Connect, Implement, and Enable*. The following diagram illustrates the path to cloud modernization:

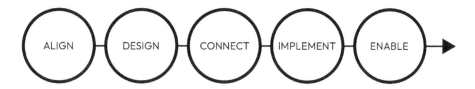

Figure 5.4 – Path to cloud modernization

Let's take a look at what each of these steps entails.

Align

There is always a risk of creating a siloed IT environment within your organization when you take up cloud projects in a standalone manner. To make sure that the outcome of the cloud model benefits your organization in terms of efficiency and flexibility, it is important to ensure your organization's business goals align with the cloud modernization goals.

Partnering with key stakeholders across the organization to ensure that all the parties are aligned and on board and agree to the implications is important while planning for modernization. Once you have drafted the plan, make sure you socialize the roadmap and ensure that your teams understand and are part of the modernization process.

Alignment usually starts with your business executives, along with the different stakeholders. Here are some of the practices that can help you align your business and the cloud:

- **Build a balanced cloud model**: Organizations typically have hybrid environments that use multiple third-party vendors to run their applications. As you evaluate your workloads for the cloud, ensure that your application stack's core requirements are taken into consideration and that you can achieve the cloud's flexible capacity and "*pay-as-you-go*" efficiencies to start with.

- **A long-term strategy for your infrastructure**: Ensure that you are advocating for flexible deployment choices to promote scalable and modular infrastructure with your cloud providers. Additionally, running interoperable applications can increase cost-effectiveness and help you adapt the cloud to your future business goals.

- **Cloud wherever possible**: When choosing to run on the cloud, ease of deployment becomes crucial for IT staff or even partners to leverage your infrastructure from any location. Ensuring that your modernization path considers easy deployment patterns will be key.

- **Avoid silos**: As discussed earlier, cloud adoption can often lead to each team having its own way of choosing computing hardware, software, and resourcing. Enterprises must develop a standardization platform and move toward consolidating their legacy applications and traditional data centers. Continuously exploring the provisioning and governance processes is key to avoiding **cloud silos**.

- **Need for governance**: As you prepare the modernization plan, ensuring a governance framework is being developed can help the enterprise gain a balanced governance model. Having insights into the usage across your cloud environment provides you with ways to look for patterns, usage trends, and anomalies so that you can take action when security breaches, data leaks, or uncontrolled costs come up promptly.

- **Need for usage and finances**: Ensure that your modernization plan also has commitments to run a cloud financial management model and that your cloud consumption is aligned with the approved forecasts. With this model, you can carry out financial analysis, where your teams can collaborate with financial professionals to make sure that the forecasted financials are aligned with your business model.

- **Involving the right people**: Involving cross-functional teams is crucial from the beginning of your modernization journey. Making sure you include the following roles can be a differentiating factor:

  - **C-level executives**: Involving leadership can ensure that you are getting the required guidance for your strategy and planning. Communicating the mission of cloud modernization to the key stakeholders is streamlined with this.

  - **Application leaders**: Application leaders can mainly provide input to your vision. These leaders can also work with their IT team leads to incorporate information across platforms while defining the modernization strategy.

  - **Infrastructure leaders**: This group of leaders can provide you with information about the required infrastructure capabilities to fulfill your modernization strategy. Members of this group are also responsible for managing the infrastructure required for the IT operational mechanisms while implementing the migration.

  - **Technical professionals**: The technical teams can provide key inputs for the architecture changes to enable modernization and are crucial while selecting the best-fit tools and services while modernizing on the cloud.

Now that you have the aforementioned aspects in place, let's take a look at the next step: design.

Design

Once the alignment is complete, it's time to develop a North Star for your modernization model. In this step, you start by examining reference architectures and evaluating options such as microservices, serverless, containers, and other key approaches.

Your plan should also entail how you plan to track the successes, failures, and lessons learned from each phase of the modernization initiative. Identifying the right tools that you want to use to adapt and solve technical problems early can help you reduce risk and realize benefits.

How to build a modernization strategy

To achieve a superior end user experience, your company's technological footing should be one of its core strengths. Let's look at the five main things you should do to build an effective IT modernization strategy:

- **Define milestones**: Having a clear vision and measurable action items with clear timelines ensures that you can translate goals into business development activities. In this phase, you can use trend analysis and benchmarking to establish additional guidelines.

- **Experiment**: As you iterate, ensure that you are testing these models at a small scale. You should test your assumptions continuously to ensure these modernized models are ready for a full-scale turnover.

- **Observe and measure**: Auditing can enable you to track change management and ensure that you are aligned with the **KPIs**. As you run through these models, ensure you have an agile mindset for productive streamlining.

- **Optimize resources**: Using your workforce and resources effectively ensures that you can monetize from capital. Use product managers to manage and ensure discipline and allocate responsibilities as per core competencies.

- **Stabilize resources**: Losing skilled resources can kill the pace of your productivity. You need tenured, skilled, and seasoned leadership to pave the way for venturing modernization efforts, and ensuring that they are ready to take a hands-on-deck approach will be critical.

After stepping into the design phase of developing a well-defined modernization model, you will need to start looking at the next step: connect.

Connect

Integrating infrastructure and connecting your data across teams to streamline enterprise business processes can help enterprises take advantage of the optimization and flexibility of cloud platforms. As part of this modernization effort, ensure that you use APIs to unlock modern architecture and digital transformation.

Over several decades, there has been an evolution in the *integration landscape*, where enterprises are slowly starting to move toward newer architectures and technologies.

There are many integration patterns, such as **point-to-point (P2P)**, **file transfer (FTP)**, **event-driven**, **SOA service**, and more. Let's look at some of the most popular patterns.

Point-to-point integration is the simplest form of integration between two or more endpoints. This is the simplest form of integration, although it can introduce technical debt if the architecture is complex and hard to manage or change.

File transfer has been around since the early stages of enterprise integration and is where an application can write a file that a second application can read and vice versa. The file format, read/write privileges, location, and standards are all tightly coupled and need to be negotiated between the applications beforehand.

The following figure depicts the evolution of the integration landscape. This is how many enterprises have evolved their integration styles per their changing needs:

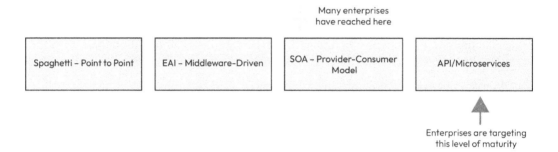

Figure 5.5 – View of integration styles

Service-oriented architectures became increasingly popular in the 2000s in the form of web services. These became a primary part of developing and implementing software applications. However, the IT industry has seen some drastic changes due to disruptive forces such as mobile, data, social networking, and the cloud. This propelled enterprises to modernize at a new level where business functionalities can be accessed via mobile applications instead of a web portal.

As a result, many enterprises target API architecture patterns to meet the sheer demand and expose business functionalities in a managed, accessible, monitored, and adaptive way through **event-driven architectures**.

It can become complex to define a modernization roadmap with the current integration landscape, so a well-thought approach can help you define a solid enterprise integration modernization plan.

The following figure breaks down the approach into three steps:

Figure 5.6 – A holistic view of enterprise integration modernization

To simplify the roadmap, here's a three-step plan that you can incorporate into your modernization plans:

1. **Assess and analyze**: Start by understanding your current integration architecture and any security or compliance requirements. Capture data about these integrations, along with the source and target information. You can then use this information to assess the various possibilities and come up with a recommended integration pattern.

2. **Envision the target state**: Once you can determine the target integration pattern and architecture from the prior step, capture the requirements that you will need to implement for the target deployment model. Use reference architectures such as event-driven architectures or microservices to gather the key considerations and create your target architecture.

3. **Define the integration portfolio roadmap**: With the target architecture model, key considerations, and list of integrations in place, ensure they are aligned with your business case. Determine how you can break the entire process into smaller logical steps to ensure proper end-to-end execution. It is always a good practice to identify a **minimum viable product** (**MVP**) and take it through implementation before starting any larger modernization efforts.

Now, let's take a look at the implementation step, which occurs once you have built your roadmap and connected the core aspects of your modernization plan.

Implement

Implementation is a crucial step in cloud modernization since it's important to keep the risk under control. To get the full value, your strategy should focus on making incremental changes and prioritizing your business cases so that the enterprise can manage modernization optimally:

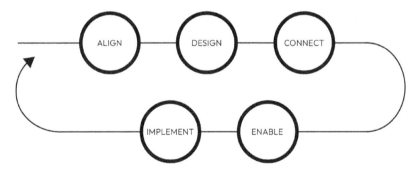

Figure 5.7 – Circular effect of the ongoing process

In this step, your implementation roadmap has already been built out and your goal should be focused on revitalizing your workloads by using the modernization mechanisms that you have already chosen to power your digital transformation.

Here are some of the categories of modernization you can implement for enterprise modernization:

- **Replatform**: This usually refers to migrating or re-hosting your current infrastructure to the cloud. The integral code base remains the same but the code is readjusted so that you can use cloud products as much as possible.

- **Remodify**: This form of modernization aims to implement new technologies so that you can achieve additional capabilities on top of the existing software layer. With this approach, your objective is to use advantages such as big data to improve customer experience, data visualization, and data discovery. Similarly, another approach for IT modernization strategies is containerization.

- **Reimprove**: This approach involves reworking the gaps and improving your existing technology stack – for example, reconciling data to simplify business processes. Additionally, system integration can be implemented to help you overcome challenges such as third-party vendors having seamless access outside the organization.

- **Replace**: This type of modernization involves substituting outdated or legacy infrastructure with state-of-the-art cloud technologies. Using this approach, you can push businesses to explore new possibilities.

- **Reconsider**: Sometimes, doing nothing could be cutting your needs and pointing toward profitability. Paving your own decisions and backing your decisions using the data that's available can help you with the dilemma of whether to modernize or not.

Now, let's look at the next step and discuss how to enable and pace your modernization efforts.

Enable and accelerate

Enabling modernization across your enterprise is a long journey and for most of you, the efforts will be ongoing. As you modernize each of your workloads, compare their behaviors and observe them closely. There could be areas where you need to fine-tune or resolve to ensure optimized results in terms of performance or cost.

These steps, if implemented in the transformation roadmap, can help you achieve successful cloud modernization. Failing to plan thoroughly can lead to failure, which you may not be prepared for. Continue to iterate and fine-tune your plans as you learn through these implementations. Some of the ways you can accelerate enterprise modernization are as follows:

- **Architecture evolution**: Your legacy footprint can often be complex and cumbersome to change. Incremental advances can help ease the process of changing the static nature of such systems. Understanding how the legacy architecture works and decoupling it from the technology helps you unlock the usage of cloud-native services. Identifying and creating product APIs can enable more business value. Creating decoupled microservices can give you the potential to rapidly innovate.

- **Modernization springboards**: Build processes that can be used repetitively and standardized so that the engineering teams can accelerate the path from idea to production. With cloud-native services, you can procure infrastructure with a click or an API call, which otherwise typically takes months. As a leader, enable your engineers to accelerate the process of bringing ideas and implementing them for new products or updates.

- **Self-service strategy**: Many enterprises rely on centralized data warehouses or data lakes for a common throughput. Taking this to the next level by delivering data as a product can make it more accessible to your customers. Easy consumption of data assets and using a self-service data strategy can help you align ownership and data management responsibilities to deliver more business value.

- **Infrastructure automation**: Removing undifferentiated heavy lifting can help free up developers so that they can focus on the business instead of managing your infrastructure. The team can step away from day-to-day operations and focus on how they can deliver business value. Infrastructure automation via code and continuous delivery increases the execution pace and reduces the risk of manually handling these processes.

In the next section, we'll discuss how enterprises can deliver both continued use and real value by moving to modern architecture.

Uncovering the stages of modernization

Many enterprises are taking a leap in legacy modernization efforts to double down on the many benefits that the cloud and modern technology can offer. Let's look at how enterprises reach these stages when it comes to their modernization journey.

Stage 1 – enabling accessibility

The rapidly changing needs of employees require enterprises to provide a framework for employees to work from anywhere without having to be restricted by location. Employees should be able to collaborate worldwide, thereby making *work from anywhere* a basic requirement. The need for collaboration platforms and videoconferencing requirements is urging leaders to rethink their cloud strategies. The availability of many SaaS tools such as Zoom, Microsoft Teams, and Slack has accelerated digital communication channel adoption and enabled companies to quickly switch to digital channels. This momentum of digital and cloud infrastructure has enabled enterprises to start switching to such tools as the first key step in their modernization journey.

Stage 2 – integrating with cloud-native

Once the accessibility challenge has been solved, the employees are set up to effectively work and improve the momentum of their efforts to transition their legacy workloads into a cloud consumption model. Many businesses find this stage tougher, given the decisions that the leaders, architects, developers, and many others involved need to make. There is no one-size-fits-all approach that you can follow but there are certainly best practices, lessons learned, case studies, and strategies that you can incorporate.

Many enterprises tend to take a portfolio approach when it comes to iteratively modernizing their organizations. The approach of redesigning applications that they have already lifted and shifted into the cloud to optimize for years to come is the main agenda for many CTOs. The business benefits compel these enterprises to modernize and enable their heartbeat applications to meet the dynamic demand.

Stage 3 – moving legacy apps to the cloud

The final stage for many enterprises is moving their most complex legacy workloads to the cloud. 20- to 39-year-old supply chains or payment systems running on mainframes written in Cobol are typical examples of such complex systems. When thinking about gaining a competitive edge, running on a mainframe will not allow you to meet current or future business demands. The real heavy lifting for such enterprises will involve going through a digital transformation where these big mainframe systems will be moved to the cloud. These workloads, being the absolute core of the business, are the first step in the direction toward modernization and open the door to a more drastic rewrite, such as breaking the monolith into microservices and building reusable APIs.

Understanding migration versus modernization

Technology leaders are looking for ways to transform their enterprises, where they begin with a lift-and-shift approach and develop a strategy to look beyond migration. For a few companies, migration is the first step within modernization, and they always go hand in hand. For a future-proof modernization effort, it is important to continually push beyond the boundaries of a lift-and-shift approach.

Looking beyond migration also means connecting data, processes, technology, and people so that they can align with the pace of change that we are experiencing in the cloud world. As a leader, you may spend a lot of time discussing whether to just **migrate** or **modernize** when it comes to transforming legacy systems. Many enterprises use the words **migration** and **modernization** interchangeably. Let's take a look at what each of these terms means.

Migration is often considered an approach to moving systems from one platform or environment to another. Enterprises choose this option for the obvious reason of the target platform being superior compared to the current platform. Betterment in terms of security, overall costs, performance, resiliency, and supportability are all preferred reasons. In the case of migration, the concept is to replicate your current systems as much as possible to avoid many changes to the functionality.

Improving part of the overall process through a migration approach can improve immediate concerns or challenges but it does not cover the long-term or big-picture vision, which involves considering an organization's growth objectives. For enterprises that are constrained by resources, migration is often a first consideration.

Modernization is defined as an approach to improving or enhancing the current capabilities of your systems. Redesigning the legacy system and upgrading it are the core criteria of modernization. Often, modernization involves designing and programming a brand-new system to replace the outdated one. Ideally, the new system should fit your current and future needs to provide the benefits of allowing you to update the infrastructure and technology footprint.

Modernizing puts technology in place to improve and integrate new and existing systems. Many enterprises lean on modernization to automate their systems end to end, improve communication, collaboration, and resource management, and enable real-time reporting:

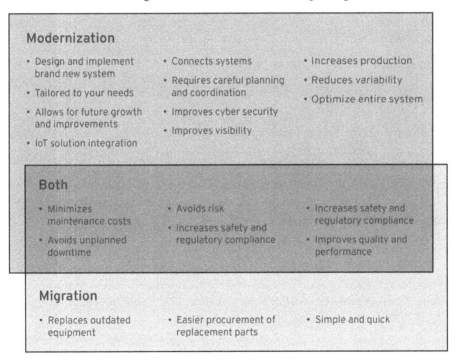

Figure 5.8 – Modernization versus migration

Choosing one of these options so that you provide the highest effect and value to your organization will create the desired effect in terms of maximizing its impact. While migration is inevitable for survival, modernization is required for your business to be competitive. Improving overall operations is not always practical for companies. However, a small-scale migration can often help prove the case for a more comprehensive modernization effort.

As discussed previously, to move your modernization efforts from planning to implementation, you must build a cohesive framework. An ideal implementation will allow you to power digital transformation and adopt a consistent, repeatable, and measurable approach. In the next section, we will discuss some of the strategies you can incorporate to modernize complex mission-critical systems at scale.

As discussed earlier, modernization is the overhaul of a production system and requires careful planning, coordination, and resources. We also discussed how choosing a phased approach and maintaining a greater vision will serve the greater overall modernization strategy. We'll learn about the business benefits of cloud-native modernization in the next section.

Exploring the benefits of modernization

Many organizations, from start-ups to enterprises, face the challenge of operating in an entirely new digital arena. These organizations' infrastructure is typically comprised of legacy systems; only a small portion of their technology is modernized. However, in the present digital world, it is important to run flexible and agile systems to overcome any hurdles and meet the emerging demands of digital business.

Let's look at the potential benefits that modernization can bring.

Competitive advantage

Modernization empowers enterprises to incorporate modern technologies and integrate them with their platforms. Digital adoption across your organization enables you to stay ahead of competitors, who may still be contemplating whether to modernize or not. For long-term business viability, organizations need to be able to respond to and excel in any unforeseen circumstances.

Paying down technical debt

Managing technical debt in organizations that are running legacy systems is the top challenge for many operations and infrastructure leaders. Through modernization, you can enable automation and self-provisioning, thereby providing a means for the technical resources to eliminate how much time they spend on maintenance tasks. With improved resource provisioning, your organizations can reduce technical debt instead of working to just keep the lights on.

Reduced business risk

With cloud-native service consumption, you are eliminating your local hardware failure possibilities. In the event of an underlying hardware failure in the cloud, well-designed and architected applications can quickly fail over and recover with minimal disruption. Any business's utmost priority is business continuity and making sure its mission-critical applications are running without much downtime. Through cloud modernization, disaster recovery can be planned and implemented, thereby combatting your business from being crippled by potential outages.

Improved end user experience

With the innovative features you gain from modernizing your legacy systems, you can redefine user-facing interfaces with leading-edge technologies and innovative features. You get to make the most of the digital platform and cloud technologies, such as artificial intelligence, machine learning, and big data to transform your IT ecosystem and stay up to date with market trends to build a strong foundation for future innovation. This helps in improving customer satisfaction, thereby enhancing your brand and company's reputation.

Robust data security

Many surveys have indicated that organizations are becoming more and more vulnerable to security breaches because of legacy systems. Their incompatibility with the current security standards and authentication methods is the main reason behind this. These can be handled by modernizing – that is, by removing obsolete and incompatible systems to strengthen the security posture of your systems:

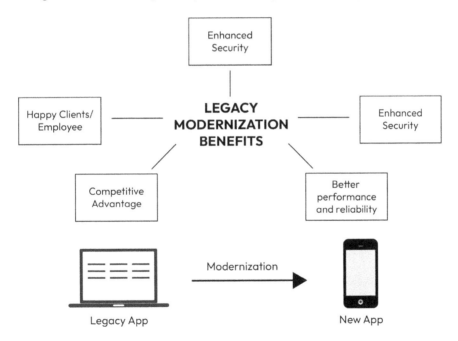

Figure 5.9 – Benefits of modernization

The preceding figure shows the benefits that we have discussed.

Cloud-native

As discussed earlier, modernization helps enterprises rearchitect their legacy systems and integrate them with cloud-native solutions. This enables you to embrace all the inherent cloud benefits, such as faster speed to market, scalability, lower costs, and better efficiency. Additionally, process-oriented operations such as updates to the systems, bug fixes, and security patches can be done easily and frequently. The operational simplicity and business performance that comes with modernization can alleviate the burden of your enterprise's IT operations.

Innovation

With modernization, you can introduce new business capabilities easily. Organizations can better handle future market and technology disruptions for any unforeseen changes in technology and customer requirements.

This agility enables organizations to be more adaptable, as shown in the following figure:

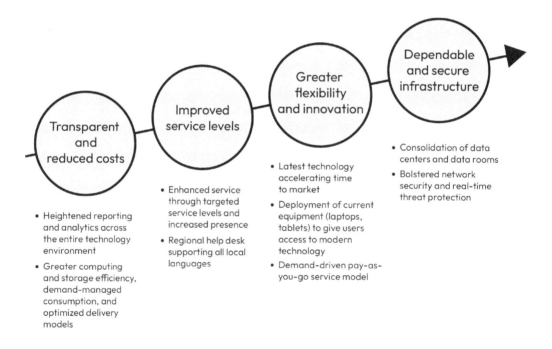

Figure 5.10 – Additional benefits of modernization

In the next section, we will introduce various AWS services and how you can use them to modernize your IT footprint. This will form the foundation for the next few chapters as we dive into each of these topics in detail.

Getting started with modernization on AWS

As mentioned in the *Uncovering the stages of modernization* section, we need to double down on stage two – that is, moving to cloud-native. For that, we need to understand the cloud-native services offered by **Amazon Web Services (AWS)**.

AWS is a cloud provider that offers a wide variety of services – more than any other cloud provider in the world. Its user base spans millions of active customers and tens of thousands of partners globally. Organizations across every industry and of every size, whether they be start-ups or large enterprises, run their workloads on AWS. Every imaginable use case can be seen running on AWS, which makes it favorable for you to bring in your requirements and adapt them easily. You can innovate more quickly with the latest technologies and accelerate the pace of modernization to transform the business at a greater scale.

With AWS, it is completely up to you how you would like to steer your modernization efforts and at what pace. Given the vast variety of services that AWS offers, you can choose what makes the most sense for your business needs and drive the most value for your investments. In the next few sub-sections, I will cover the fundamental technology domains that are available within AWS so that you have a comprehensive list of resources at your fingertips.

Fundamental technology categories

To get the most out of the AWS cloud platform, you need to understand the available offerings and the use cases that it is going to solve. There is an exhaustive list of services that you can find on the AWS website. This section does not intend to duplicate that effort or provide a complete list of the offerings but to provide you with a foundational mental map of the core technologies that we will be diving deep into in the next few chapters that are key for digital transformation and modernizing any enterprise.

As we cover these AWS services, you will notice that the main focus on digital transformation and modernization is to remove the undifferentiated heavy lifting of procuring hardware and infrastructure maintenance so that you get to shift focus to your business. AWS has an ecosystem of solutions and services that are designed to handle this. We will introduce some of them here, along with the foundational topics.

Compute

AWS provides the broadest and deepest set of services within the compute platform for millions of organizations to run their diverse workloads. You can manage your infrastructure with granular control and pick from multiple choices of processors, storage, and networking. When it comes to leveraging cloud compute services on AWS, there are some core services you can use.

Amazon **Elastic Compute Cloud** (**EC2**) provides computing capacity in the AWS cloud. With Amazon EC2, you get scalable infrastructure that you can use to develop and deploy applications. You can procure instance types that can be used for general-purpose, compute-optimized, memory-optimized, storage-optimized, or accelerated computing to give you the most optimal experience when it comes to compute, memory, storage, and networking for your workloads. The processors for these instance types are powered by Intel, AMD, NVIDIA, and AWS, which further enhance the performance and help in optimizing the costs.

For applications that need direct access to the processor or memory for the underlying server to run non-virtualized environments, you can use bare-metal instances. You can launch as many EC2 instances as you need within minutes and eliminate the need to invest in hardware upfront. Technology such as the AWS Nitro system enables a combination of dedicated hardware and a lightweight hypervisor to power faster innovation and enhanced security.

Here are some of the key features of Amazon EC2:

- The use of instances to procure virtual computing environments.

- Pre-configured templates to bootstrap the instances.

- The ability to spin up your virtual machine with limited friction.

- A secure compute for your applications. Security is built into the foundation of Amazon EC2 with the AWS Nitro system (`https://aws.amazon.com/ec2/nitro/`)

- The option to configure CPU, memory, storage, and network capacity.

- The ability to log in securely using key pairs, a combination of a public and a private key.

- The ability to choose and use storage volumes that are temporary as well as persistent.

Recently announced, Amazon EC2 Mac instances allow you to run on-demand macOS workloads in the cloud, thereby extending the flexibility, scalability, and cost benefits of AWS to all Apple developers. With Amazon EC2, you get a committed SLA of 99.99% availability so that you can run your mission-critical applications and meet high-demand business needs.

Amazon EC2 Auto Scaling is a capability that helps you manage your application's availability by adding or removing EC2 instances automatically according to the conditions you configure. Autoscaling's fleet management feature lets you maintain the health and availability of your instance fleet. With fleet management, you can detect impaired EC2 instances and unhealthy applications and replace them with healthy instances without any intervention. This way, the application's compute capacity is ensured at any given point in time.

Additionally, the dynamic and predictive scaling features let you add or remove EC2 instances to automatically respond to changing demand. Predictive scaling automatically schedules the right number of EC2 instances to be active based on the predicted demand. Predictive scaling uses machine learning algorithms to detect changes in daily and weekly patterns and adjust the compute forecasting accordingly.

Here are some of the key features of EC2 autoscaling:

- The ability to monitor the health of running instances
- Replace impaired instances automatically
- Balance capacity across availability zones
- Scheduled scaling
- Protected by AWS global network security procedures and within the scope of compliance programs such as SOC, PCI DSS, FedRAMP, the HIPAA, and others
- Dynamic scaling to follow the demand curve for your applications
- Predictive scaling to schedule the right number of EC2 compute instances
- Overall faster, simpler, and more accurate capacity provisioning without manual intervention

Amazon EC2 Autoscaling is considered one of the best practices to mitigate both infrastructure- and application-layer attacks. If you are running web applications, you can use load balancers to distribute traffic to several Amazon EC2 instances that are overprovisioned or configured to automatically scale. You can scale horizontally by automatically adding instances to your application by scaling the size of your Amazon EC2 Autoscaling group.

Autoscaling has many important benefits when it comes to safeguarding your EC2 instances and you can achieve payoffs both in terms of the operational efficiency and the reliability of the application. This is a crucial concept for services such as AWS Elastic Beanstalk and AWS CodeDeploy for fine-grained control over the deployment processes.

Amazon Elastic Container Service (**Amazon ECS**) is a fully managed container orchestration service that lets you run highly secure, reliable, and scalable containers on AWS. With its deep integration with the rest of the AWS platform, you get a secure and easy-to-use orchestration service for running your container workloads on AWS. ECS comes with the powerful simplicity to enable you to grow from a single Docker container to managing an entire enterprise application portfolio.

Here are some of the key features of ECS:

- Docker support to run and manage Docker containers.
- Integrated with Elastic Load Balancing, allowing you to distribute traffic across your containers.
- Native integration with AWS's security, identity, management, and governance tools.
- A fully managed container orchestration service, with AWS configuration and operational best practices.
- You can automatically recover unhealthy containers to ensure that you have the desired number of containers to support your applications.

- The control plane is fully managed to deliver a secure and reliable service. AWS provides out-of-the-box security configuration to protect the container images from vulnerability and/or misconfigurations.

- Supports batch processing and can scale web applications in multiple **Availability Zones (AZs)**.

With ECS, you can move forward with the existing solution rather than building one from the ground up. Reallocating resources from infrastructure management to product development becomes easy using ECS.

Amazon **Elastic Kubernetes Service (EKS)** is a managed container service where you can run and scale Kubernetes applications in the cloud or on-premises. Kubernetes is an open source container orchestration system for automating the process of deploying, scaling, and managing containerized applications. Your existing applications that run on upstream Kubernetes are compatible with Amazon EKS, given that EKS is a certified Kubernetes-conformant.

EKS lets you take advantage of the performance, scale, reliability, and availability of your AWS infrastructure.

Here are some of the key features of EKS:

- Can deploy across hybrid environments

- Can model machine learning workflows to efficiently run distributed training jobs

- Can build and run web applications in a highly available configuration across multiple AZs and use out-of-the-box networking and security integrations

- Enables service integrations to give you direct management control over AWS services from within your Kubernetes environment

Many enterprises use EKS as a strategic direction for migrating their workloads. You can run over millions of transactions per second through EKS and realize a significant reduction of developer effort while launching microservices. With its unprecedentedly high availability across the globe and its affordability, many enterprises have shifted focus and run their complex workloads on EKS.

Amazon Lightsail offers you cloud resources so that you can build your web application and get it up and running in just a few clicks. Resources such as instances, containers, databases, storage, load balancers, DNS management, and more are good examples. Spinning up websites from blueprints such as WordPress, Prestashop, and LAMP is possible with Lightsail.

Using Lightsail, many small businesses, students, or other users can build and host their applications on the cloud with minimal effort. You can choose from pre-configured instances that are pre-baked with everything you need to deploy and manage your application. Compute, storage, and networking capacity can be provided to developers and used to deploy and manage websites or web applications. With Lightsail, you can easily procure the resources to launch your project quickly for a low monthly price.

Here are some of the key features of Amazon Lightsail:

- You can click to launch a simple operating system, a pre-configured application, or deployment stacks through virtual servers or instances that are backed by the power and reliability of AWS

- You can easily run containers in the cloud from the Docker images pushed by developers

- Simplified load balancing routes so that your websites and applications can meet the variations in traffic and stay protected against outages

- Fully configured MySQL or PostgreSQL databases

- You can procure both block and object storage with highly available SSD-backed storage for Linux or Windows virtual servers

- You can enable CDN distributions to distribute your content to a global audience

With Lightsail, you can innovate quickly and cost-effectively, even with smaller teams, while managing your business environment. You can host your applications on AWS easily and provide your end users with low latency and global availability.

AWS Lambda is a serverless compute service that's suitable for event-driven architectures. Serverless empowers end users to build and run their code without worrying about the underlying infrastructure. Compute resources such as servers or clusters will automatically be facilitated to run and support your applications. You can write custom logic and leverage AWS's ecosystem to run your applications at scale. Performance, reliability, and security are guaranteed with the AWS services that you can leverage in conjunction with Lambda.

Enterprises of any scale and size can run interactive web and mobile backends by combining AWS Lambda with other AWS services to provide secure, stable, and scalable end user experiences. You can preprocess data using AWS Lambda before feeding it to your machine learning model. Building event-driven functions for easy communication between decoupled services becomes much easier with Lambda.

Here are some of the key features of Lambda:

- With no servers to manage, Lambda provides built-in logging and monitoring, including integration with Amazon CloudWatch, CloudWatch logs, and AWS CloudTrail

- You can write your code in programming languages such as **Java**, **Go**, **PowerShell**, **Node.js**, **C#**, and **Python**

- You can run highly available and fault-tolerant applications with fully automated administration

- You can package and deploy container images for Lambda-based applications with the container image tooling

- You can interact with relational databases to build secure Lambda-based serverless applications

- As a managed service, AWS Lambda reduces the attack surface while making cloud security simpler

- You can connect securely and read, write, and persist large volumes of data to shared file systems such as Amazon **Elastic File System (EFS)**, at low latency, at any scale

- You can connect to Amazon's CDN, CloudFront, and respond to events that run globally with low latencies

There are many use cases for serverless architectures, mainly web systems, the **Internet of Things (IoT)**, connecting to microservices via APIs, automation, and many more. Many enterprises use AWS Lambda to adopt serverless architecture-based solutions for core systems. We will dive deep into the various use cases, best practices, recommendations, and much more concerning containers and serverless in upcoming chapters.

Security

AWS treats security as job zero, where the infrastructure has been architected and built to be the most secure cloud computing environment. The security best practices and standards have been incorporated into its infrastructure. AWS uses redundant and layered controls, does continuous validation and testing, and incorporates a substantial amount of automation to ensure 24/7 monitoring and protection.

You can achieve compliance with most security-sensitive requirements by leveraging AWS's ecosystem, which meets the standards of resilient infrastructure and high-security protocols. AWS recommends its shared responsibility model, where AWS takes ownership of the security of the underlying infrastructure, and you take ownership of securing your applications. By following the shared responsibility model, you get the flexibility and agility to leverage the security controls and run your business functions with restricted access to the public:

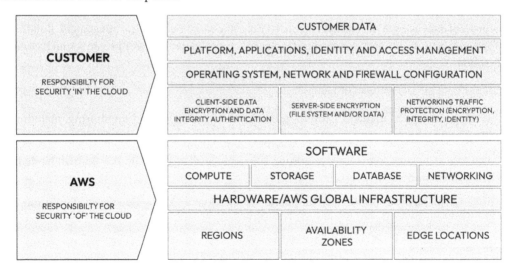

Figure 5.11 – AWS Shared Responsibility Model

AWS provides security capabilities such as network firewalls that are built into Amazon **Virtual Private Cloud** (**VPC**), a virtual network for launching AWS resources. You get to control encryption in transit with TLS across AWS services. You also get to choose from several connectivity options such as private, dedicated, or from your office or on-premises environment. Automatic encryption of all traffic is applied by default between the AWS-secured facilities. DDoS mitigation technologies are applied at layers 3, 4, and 7.

AWS Identity and Access Management (**IAM**) lets you define individual user accounts with permissions across AWS resources for the accounts that you configure on AWS. IAM can further be used to control access to your company's employees and applications via AWS. You get fine-grained access control across all AWS resources.

Here are some of the key features of IAM:

- Fine-grained access control to AWS resources such as Amazon EC2, **Amazon Simple Storage Service** (**Amazon S3**), Amazon DynamoDB, and others

- You can delegate access to users or AWS services that operate within your AWS account by using IAM roles

- You can use IAM Access Analyzer to follow the principle of least privilege and grant the right fine-grained permissions to meet your evolving requirements

- You can apply permission guardrails for all accounts to adhere to

- It provides attribute-based access control so that you can create fine-grained permissions based on user attributes such as job, role, team name, and so on

With IAM, you get to manage the permissions of your enterprise's workforce and establish the required permission guardrails and data perimeters across your AWS organization.

AWS Single Sign-On (**AWS SSO**) allows you to manage central access and connect to your workforce in AWS. AWS also lets you bring your own sign-on providers, such as Microsoft Active Directory or OKTA, so that you can create user identities in AWS SSO. SSO provides a unified administration experience where you can define, customize, and assign fine-grained access.

Here are some of the key features of SSO:

- Enable SSO access for multiple AWS accounts centrally

- Manage access to multiple AWS accounts from one place

- Manage access to your cloud applications

- Connect to Microsoft Active Directory

- Audit SSO activity across applications and your AWS accounts

- Built-in SSO integrations with business applications

- Automatically provision users from standards-based identity providers

With SSO, you get pre-configured settings for many cloud applications, such as Salesforce, Box, Microsoft 365, and others. By deploying AWS SSO on your accounts, you get to use your existing corporate credentials without the hassle of managing a traditional SSO solution. Easily and quickly connecting users to AWS using your normal enterprise credentials lets you focus on continuing to deliver services to your customers instead of managing user credentials in a multi-account structure.

Analytics and big data

With the world becoming more and more digital, there is exponential growth in the amount of data that is being created and collected. Traditional analytical tools have their challenges when it comes to analyzing this ever-growing data. Real-time and batch processing of structured and unstructured data over high-velocity transactions on complex architectures are the evolutions that data management architectures have witnessed.

When it comes to modernizing your IT footprint, transforming your data storage, collection, and analysis aspects becomes the core of how you provide improved customer experiences and gain a competitive advantage in the market to grow your business as a leader. AWS has a wide variety of managed services to help you build, secure, and seamlessly scale end-to-end big data applications quickly and easily. With these services, you can perform real-time streaming or batch data processing efficiently and tackle your next big data transformation projects with more ease.

Amazon S3 is an enterprise-grade object storage service. S3 offers scalability, data availability, security, and performance to cater to industry-level operations. S3 can be used by companies of all sizes and industries of all types, including data lakes, cloud-native applications, and mobile applications. Data is stored as objects called **buckets**, and a single object can be up to 5 terabytes in size.

S3's flat and non-hierarchical structure makes it easy to organize data in ways that are valuable to your teams. Using this service, you can easily build applications that depend on cloud-native storage. You pay for what you use, so you can start small and grow as your business needs evolve without compromising on performance, security, or reliability. Due to S3's highly flexible nature, you can store any type of data you want, read the same object a million times, or use it for emergency disaster purposes (this is rare). Use cases range from building a simple FTP application to a sophisticated web application with S3.

Here are some of the key features of S3:

- Storage management capabilities such as data version control, metadata tagging, and replication techniques

- Storage management to monitor and control your S3 resources

- S3 storage lens for organization-wide visibility into object storage usage and activity trends to improve costs and apply data protection best practices

- S3 storage class analysis to help you decide when to transition the right data into the right class for cost-effectiveness

- Access management to have fine-grained control and simplify management of the data access to shared users or resources

- Security controls to block unauthorized users and mechanisms such as client-side encryption and server-side encryption for data uploads

- An S3 Lambda object for you to add custom code to modify and process data coming from S3 to an application

- Built-in capability to query data without needing to copy and load it into separate analytics platforms or data warehouses

- Supports parallel requests, which means that you can scale your S3 performance by the factor of your compute cluster without making additional customizations to your applications

- Strong read-after-write consistency without sacrificing regional isolation for applications at no additional cost

With S3, many enterprises can easily support a high volume of transactions easily and support the growth of their data volumes and complexity. S3 can easily become a cornerstone of your company's data infrastructure, whether it comes to building a data lake or simply storing volumes of historical data that need high durability. S3 allows you to act on data that was previously unactionable and can be centrally managed and operated easily, ultimately improving the end user's experience with guaranteed global manageability.

AWS Glue is a serverless data integration service that you can use to discover, prepare, and combine data for analytics, machine learning, and application development. You get the capabilities you need for data integration in minutes, which you would otherwise spend months getting using traditional analytical products. Typically, different products are used to enrich, clean, normalize, combine, and load/organize data in data lakes or warehouses. With Glue, you can do all of this using visual or code-based interfaces with a few clicks.

Here are some of the key features of Glue:

- Ability to discover and search across all your AWS datasets to compute statistics and gather additional insights

- Automatic schema discovery to ensure your metadata is up to date

- Schema registry to validate and control the evolution of streaming data

- Visually transform data with a drag-and-drop interface so you can author highly scalable **extract, transform, load** (ETL) jobs for distributed processing

- Build complex ETL pipelines on a schedule, on-demand, or event-based in parallel

- Create unified catalogs to make it easier to search for data across multiple data stores

- No coding is required to create, run, and monitor ETL jobs

- Create materialized views for combining and replicating data

- Experiment with data from data lakes, data warehouses, databases, and Amazon S3

Glue enables you to extract, transform and load from your data lakes and power the processes to transform your data while you pay only for the computing power that you need to run these jobs. Data analysts and data scientists get more control to enrich data without having to write code. With the 99.9% SLA guarantee, you get to interactively develop your ETL code and normalize data easily.

Amazon Redshift is a fully managed and scalable data warehouse service. It is primarily used for fast, easy, and secure analytics at scale. Redshift offers a platform where you can analyze terabytes to petabytes of data so that you can run complex analytical queries easily. You can capture real-time insights and perform predictive analytics on data across multiple operational databases, data lakes, data warehouses, and third-party datasets.

Given its ease of use and performance at scale, Redshift becomes an obvious choice for many enterprises to run and scale their analytics without having to worry about managing a data warehouse. Redshift supports industry-leading security standards and is compliant with SOC1, SOC2, SOC3, and PCI DSS Level 1 requirements.

Here are some of the key features of Redshift:

- Ability to get insights from data in seconds without worrying about managing a data warehouse

- Capture integrated insights and run real-time or predictive analytics on complex, scaled data across databases, data lakes, data warehouses, and thousands of third-party datasets

- High performance per price than other cloud data warehouses and automated optimizations to improve query speed

- Comprehensive security capabilities such as end-to-end encryption, network isolation, audit and compliance, and tokenization to align with the most demanding requirements of enterprises at no extra cost

When it comes to migrating from an on-premises data warehouse to a high-performing data hosting solution, Amazon Redshift can be strongly considered given its deep integration with the AWS ecosystem and shortened ETL times.

Amazon Athena enables data analysts to perform interactive queries and analyze data in Amazon S3 using Standard SQL. This is a serverless service, so you pay for the queries you run and don't need to manage infrastructure. Athena lets you query data without having to set up and manage servers or data warehouses. Data analysts can define their schema and point to the data in Amazon S3 to start querying using the built-in query editor. You can tap into your data without setting up complex processes to ETL the data.

Here are some of the key features of Athena:

- You can run your queries on a serverless platform, so there's zero infrastructure and zero administration.

- Interactive performance for datasets of any size and you get parallel execution of queries so that you can capture results within seconds.

- Athena runs on Presto, an open source distributed SQL query engine optimized for low latency that supports ad hoc analysis of data.

- Parallel query execution and supports fast performance with Amazon S3.

- Integrates with IAM to control access to your data.

- Integrates out of the box with AWS Glue to create a unified metadata repository across various services and crawl data sources to populate your Data Catalog.

- Ability to invoke machine learning models in an Athena SQL query for tasks such as anomaly detection, customer cohort analysis, and sales predictions through a simple SQL query.

Query services such as Athena address different needs and use cases to run ad hoc queries for data in S3 without the need to set up or manage any servers. You can use Athena if your data catalog is a Hive metastore and have the need to use federated queries to run SQL queries across a variety of relational, non-relational, and custom data sources.

Databases

AWS provides more than 15 purpose-built database services to serve enterprises, from a range of open source to enterprise-grade commercial databases. Merging flexibility and low cost, AWS databases are designed to meet the demands of modern and globally distributed applications running microservice architectures.

You get to choose a database service that fits your use case and suits your specific business needs. The choices vary and support data models such as relational, key-value, document, in-memory, graph, time-series, wide column, and ledger databases, as shown in the following table:

Database Type	Use Cases
Relational	Used for traditional applications, **enterprise resource planning** (ERP), **customer relationship management** (CRM), and e-commerce
Key-value	Serves high-traffic web, e-commerce, and gaming applications
In-memory	Used for caching, session management, gaming leaderboards, and geospatial applications
Document	Serves content management, catalogs, and user profiles
Wide column	Used for high-scale industrial applications for equipment maintenance, fleet management, and route optimization

Database Type	Use Cases
Graph	Used for fraud detection, social networking, and recommendation engines
Time series	Serves **Internet of Things** (IoT) applications, DevOps, and industrial telemetry
Ledger	Used for systems of record, supply chain, registration, and banking transactions

Table 5.1 – Supported database types in AWS

Amazon Relational Database Service (**RDS**) is a managed relational database that provides you with the choice of six database engine options: Amazon Aurora, MySQL, MariaDB, Oracle, Microsoft SQL Server, and PostgreSQL.

The following are some of the key features of RDS:

- Ability to administer easily without infrastructure provisioning or database software maintenance.

- Scalability to launch one or more read replicas and offload the read traffic from your primary database instance.

- Reliability to provision a multi-AZ DB instance, and synchronously replicate the data to a standby instance in a different AZ.

- High performance between multiple SSD-backed storage options, one for high-performance **online transaction processing** (**OLTP**) applications, and the other for cost-effective general-purpose use.

- Highly secured, where you can run database instances in Amazon VPC and isolate your database instances. You can connect to your existing IT infrastructure through industry-standard encrypted IPsec VPNs.

With RDS, you can run your databases with a lean cost structure, where your resources are devoted to building your businesses instead of maintenance and operational heavy lifting.

Amazon DynamoDB is a fully managed, serverless, key-value NoSQL database. With DynamoDB, you can run high-performance applications at any scale. DynamoDB offers built-in security, continuous backups, automated multi-region replication, in-memory caching, and data export tools.

The following are some of the key features of DynamoDB:

- Provides a flexible schema for storing data in key-value pairs that can easily adapt per your business requirements without you having to change any schemas. With this, you can achieve fast read performance with microsecond latency, as well as automated global replication with global tables and run globally distributed workloads.

- Includes support for **Atomicity, Consistency, Isolation, and Durability** (**ACID**) transactions to run mission-critical workloads that require complex business logic.

- Supports encryption at rest by default and further enhances your data security by using encryption keys.

- Provides point-in-time recovery and protects your DynamoDB tables from accidental write or delete operations.

- Support for on-demand backups and restores to create full backups of your tables for data archiving.

Using DynamoDB, enterprises can scale nearly infinitely with no performance issues, and you get to focus on continued innovation without having to worry about components such as replication, backups, or capacity management.

Amazon ElastiCache is a fully managed, in-memory caching service designed to support flexible, real-time use cases. With ElastiCache, you can improve application and database performance, or even use it as a primary data store that doesn't require durability.

Here are some of the key features of ElastiCache:

- Compatibility with two engines, Redis and Memcached

- Supports detailed monitoring statistics via Amazon CloudWatch

- Can cache data to reduce pressure on the backend database, thereby enabling higher application scalability and reducing the operational burden

- Stores non-durable datasets in memory and supports real-time applications with microsecond latency

ElastiCache data tiering allows enterprises to grow with a virtually unnoticeable performance impact. You don't have to rethink the underlying infrastructure to make this happen, which eases the burden for any organization running in the technology space.

Machine learning

AWS provides the broadest and deepest set of machine learning services to support cloud infrastructure, thereby making it easy for every developer and data scientist to accelerate their machine learning journey. While machine learning is one of the most exciting technologies, it is strenuous to focus on data and build and train models without proper platforms.

Amazon SageMaker is a fully managed service that enables developers and data scientists to easily build, train, and deploy machine learning models at scale. SageMaker removes the complexity that usually holds back developers so that they can deliver the best possible predictions. The heavy lifting from each step of the machine learning process can be removed and high-quality models with more precision can be built using SageMaker.

You get all the components that are used for machine learning in a single toolset so that they get to production faster with much less effort and at a low cost.

Here are some key features of SageMaker:

- Organize data in minutes using Amazon SageMaker Data Wrangler.

- Improve model quality through bias detection during data preparation and training using SageMaker Clarify.

- Operate on a fully secure machine learning environment. Here, you can use a comprehensive set of security features to help support a broad range of industry regulations.

- Get accurate training datasets using SageMaker Ground Truth Plus without having to build labeling applications or manage labeling workforces.

- Ability to extend the ease, scalability, and reliability of SageMaker to run data processing workloads using SageMaker Processing.

- No-code machine learning using Canvas, which is a visual, point-and-click service that allows business analysts to build machine learning models and generate accurate predictions.

- One-click Jupyter Notebooks; the underlying compute resources are fully elastic.

- Offers more than 15 built-in algorithms that come in pre-built container images for you to quickly train and run inferences.

With Amazon SageMaker, you can accelerate artificial intelligence initiatives and advance your business cases. You get to build and deploy algorithms on the AWS platform and solve complex problems to accelerate and power prosperity for your end users.

Amazon Comprehend is a **natural language processing** (**NLP**) service that uses machine learning to capture insights from text. You get functionalities such as custom entity recognition, custom classification, key phrase extraction, sentiment analysis, and entity recognition that can be easily integrated into your applications.

Here are some of the key features of Comprehend:

- Captures insights from text in different types of documents, such as customer support tickets, product reviews, emails, and social media feeds

- Makes document processing workflows easy by extracting text, key phrases, topics, sentiment, and more from documents such as insurance claims

- Trains a model to classify documents and identify terms with no machine learning experience

- Protects and controls access to your sensitive data by identifying and redacting **personally identifiable information** (**PII**) from documents

Enterprises can use Amazon Comprehend to automate costly manual processes such as document reviews, insurance application intake, identifying supply chain risk, redaction within a tokenization system, improving identification of best-matched catalog items, analyzing and transcribing text from video content, making chat apps smarter, sentiment analysis, and much more easily and effectively.

In the next section, we will look at a case study of how an enterprise leverages solutions with a modern, future-proof architecture that is cloud-enabled and has evolved to cater to its changing business needs.

Case study

In this case study, we will explore how cloud modernization enabled a postal operator service company to reduce labor and operational costs while meeting business needs.

Current state – the challenge

A mid-sized leading postal operator, *PostABC*, is a growing parcel partner in areas such as North America, Asia, and Europe. *PostABC* is faced with the challenge of making its business digital-ready with high scalability and services available 24/7. The company's IT legacy platforms were monolithic and over 20 years old. Maintaining the applications on these platforms increased operational costs and required manual efforts for deploying new features and enhancements. The lack of trained personnel maintaining the data centers forced it to reimagine its business and migrate systems to the cloud. It was important for it to modernize its aging technology platforms. For that, it wanted to retire its mainframe and modernize it core legacy system to improve agility and responsiveness to deliver a better customer experience.

To drive more business value and growth, it wanted to overcome the cost involved in operating on-premises infrastructure and vendor-specific software stacks. *PostABC* looked to use the cloud for business innovation and market differentiation while increasing efficiency and reducing operational costs.

AWS solution

A phased cloud modernization plan was developed to capitalize on the benefits of the cloud's elastic nature while modernizing its IT footprint. Starting with the technology stack, refactoring and redesigning the monolith legacy system into microservices was recommended. To improve elasticity and scalability, it was recommended to decouple individual solution components using AWS Managed Services. In addition, a microservice solution where each piece could easily be modified would need to effectively utilize AWS cloud-native services.

Further, these microservices were deployed by *PostABC* in containers and orchestrated using Amazon EKS. The software licenses were consolidated and replaced expensive technologies with cloud-native alternatives such as Amazon Aurora; the existing storage was replaced with Amazon S3. After that, Amazon **Managed Streaming for Apace Kafka** (**MSK**) was recommended to ensure these distributed microservices communicated effectively.

Using Amazon EKS to run containers allowed the team to configure and use the most effective type of EC2 instance. The Graviton instance family was used for these nodes inside the same EKS cluster, which improved the system's overall performance while reducing the overall cloud expenditure. Helm file configuration was used to allow the seamless deployment of services in Amazon EKS.

Additionally, using the GitFlow approach for application development and designing and building deployment environments and CI/CD pipelines to make frequent microservice application releases was recommended. The team further re-architected the legacy application and built a new CI/CD microservices testing solution. They leveraged blue/green deployment to test the cloud-native solution before migrating at scale.

Finally, to monitor all the services deployed in EKS, CloudWatch Container Insights was recommended. Experts on the team configured DeamonSet with FluentD running on nodes for advanced parsing. FluentBit is an open source lightweight process that aggregates and forwards logs, parses them, and sends the metadata to the Amazon OpenSearch service. This helps developers view the logs using pre-baked dashboards in Kibana.

Realized benefits

As a result, *PostABC* was able to move toward establishing a highly resilient and modern technology foundation to support a digitally scalable and ready business. The platform delivered 100% application availability during a record peak period. In the process, *PostABC* was also able to realize infrastructure savings of 1 million USD per year and an additional 600,000 USD in savings due to a reduction in database license costs. The calibration times were reduced up to 90%, which allowed it to provide near real-time responses to its end users. In turn, this increased the number of successful customer transactions.

The following figure illustrates all the extensive benefits that *PostABC* was able to achieve through modernization:

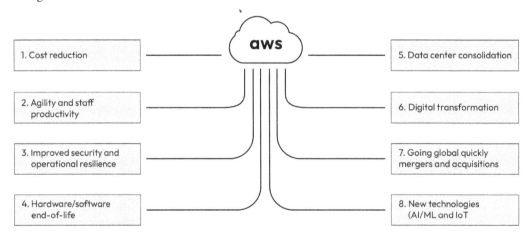

Figure 5.12 – Cloud modernization benefits

PostABC's mainframe application's migration and modernization on AWS have enabled the business agility that was needed to deploy and scale new offerings and provide an enriched customer experience. By modernizing its footprint, *PostABC* was able to enable its growth into other geographic markets.

Summary

In this chapter, we discussed what cloud modernization entails and the key strategies to incorporate and achieve cloud modernization. We also learned how migration is different from cloud modernization and some of the parameters to keep in mind while evaluating the best approach for your enterprises. Then, we discussed various foundational AWS services that are important for you to get started with before you begin cloud migration and modernization. Finally, we dove deep into a case study where an enterprise went through an end-to-end modernization journey to transform its business. In the next chapter, we will dive deep into what application modernization is and how that can be implemented using AWS services.

Further reading

To learn more about the topics that were covered in this chapter, take a look at the following resources:

- 2021 Mainframe Modernization Business Barometer Report - `https://modernsystems.oneadvanced.com/globalassets/modern-systems-assets/resources/reports/advanced_mainframe_report_2021.pdf`

- AWS Prescriptive Guidance - `https://aws.amazon.com/prescriptive-guidance/`

- Evaluating modernization readiness for applications in the AWS Cloud - `https://docs.aws.amazon.com/prescriptive-guidance/latest/modernization-assessing-applications/welcome.html`

- AWS Architecture Center - `https://aws.amazon.com/architecture/`

- Modernize Your Applications, Drive Growth and Reduce TCO - `https://aws.amazon.com/enterprise/modernization/`

- Amazon EC2 Auto Scaling - `https://aws.amazon.com/ec2/autoscaling/`

- Amazon Lightsail - `https://aws.amazon.com/lightsail/`

- AWS Cloud Databases - `https://aws.amazon.com/products/databases/`

- Amazon ElastiCache - `https://aws.amazon.com/elasticache/`

- Machine Learning (ML) and Artificial Intelligence (AI) - `https://docs.aws.amazon.com/whitepapers/latest/aws-overview/machine-learning.html`

6

Application Modernization Approaches

Gartner, Inc forecasts that in 2023, worldwide public cloud spending will grow 20.7% to a total of $591.8 billion, up from $490.3 billion in 2022. With scalability and faster time to market becoming the top priorities of businesses in today's market, **FinOps** is a popular cloud computing trend for 2023. There is close monitoring of handling costs due to macroeconomic conditions and inflation. FinOps is a practice within many companies for managing IT costs. Cloud computing and digital transformation empower you to manage your computing costs effectively and innovate to compete in the next generation influenced by technology. Modern application approaches enable you to build and manage applications that can increase the agility of your teams and the reliability, security, and scalability of your systems. The rapid pace of experimentation can be easily facilitated and becomes a fundamental shift in how you approach business value creation.

Digital transformation is redefining economies around the world, and incorporating modern technologies to solve real-life problems and day-to-day business operations is becoming the norm.

This chapter will focus on the following key topics:

- An introduction to **application modernization** (**AppMod**)
- Understanding the key strategies for AppMod
- Breaking monolithic applications into microservices
- Best practices for modern application development
- The **Amazon Web Services** (**AWS**) landscape for AppMod
- Case study

An introduction to application modernization (AppMod)

Before we start discussing what AppMod is and the right strategies to modernize your applications, often termed **legacy systems**, let us understand what legacy systems are and how to identify whether you are running them.

Legacy systems

A legacy system is an outdated system, technology, or software application that is used by an organization, which relies on its capabilities and services. Legacy systems can be the backbone applications that are holding your enterprise back from using new digital technologies – such as the **cloud**, **Internet of things** (**IoT**), or **mobile**. And this can be a deterrent to giving your customers and enterprises a modern experience.

Typically, as technology advances, these legacy systems do not have support or maintenance and are limited in terms of growth. While a legacy application may be an integral part of the company and may continue to be used, it comes with a whole host of challenges as a result of the inability to keep up with the changes in technology, architecture, or functionality. It impedes the business (value, agility, and fit) and IT (in terms of cost, complexity, or risk).

Common challenges of legacy systems

As businesses evolve, it is expected that the systems may quickly become outdated but still be in a condition where they perform the required operations and maintain the status quo. However, if you are facing any/all of the following challenges with your current systems, then treat that as an indicator that your applications are becoming obsolete:

Challenge type	Description of the challenge
Maintenance/updates	Is your system difficult to maintain or update?
Increased costs	Is your system maintenance resulting in increased IT costs in terms of both time and staff?
Ongoing operational issues	Does your system often encounter performance issues, instability, or bottlenecks in scaling?
Lack of technical specifications	Is your code undocumented or difficult to understand for new IT staff?
Incompatibilities	Is your system dependent on third-party elements that may no longer be supported?
Interoperability	Can your application no longer interact with newer systems?
Agility	Is your system no longer agile to emerging customer needs?
Security or compliance risks	Does the system have potential security or compliance risks?
Integration	Is your current application able to easily integrate with new systems?
End of life	Does your system run on dependencies that have reached end-of-life?

Table 6.1 – Legacy system challenges

You can use the preceding indicators to evaluate and analyze to what extent you are willing to maintain them in their as-is state. These issues typically cost not just your time and money but also probably hinder the chance for growth and innovation.

While these applications can be the most mission-critical ones in your business, transforming them can improve how your enterprise runs its software platforms, tools, architectures, libraries, or frameworks.

What is AppMod?

AppMod is the process of extending the lifespan of the applications while transforming traditional applications to use the **next-generation (NextGen)** framework and incorporate modern computing approaches to take advantage of improvements in terms of ongoing costs, increase agility, innovate, or get to market faster with rapid iteration and experimentation.

The following diagram shows the modernization of an application from the perspective of incorporating different technologies and approaches:

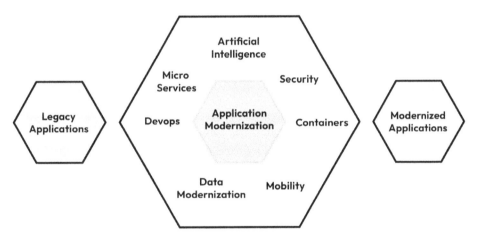

Figure 6.1 – AppMod

AppMod enables you to refresh the application portfolio and helps you avoid common pitfalls to return a more favorable **return of investment (ROI)**. While AppMod was once an IT-driven approach, it is becoming a norm that C-level executives consciously make modernization decisions to leverage the characteristics of the next generation of modern technology. There is always a question of how NextGen applications should look or what characteristics define a NextGen modern application. We will cover this in detail in the next section.

Characteristics of modern NextGen applications

There are key characteristics of the next generation of modern applications that you want to achieve while building them. These can be foundational guidelines that provide a good structure to your applications. As technology leaders, you will find the following aspects to be the key differentiators and disruptors.

Well architected

Your applications are built and benchmarked based on effective architectures to scale your business and maximize efficiency. Benchmarking allows your production workload to be stable, secure, and efficient both in terms of performance and cost.

Applications that are well architected on the cloud leverage AWS's Well-Architected Framework design principles, which are organized int six pillars: *Operational Excellence*, *Security*, *Reliability*, *Performance Efficiency*, *Cost Optimization*, and *Sustainability*. All of these are important when designing cloud-native applications. Let us take a quick look at each of these pillars:

- **Operational Excellence**: This pillar includes the ability to support and run applications effectively with insights into their operation and continuously improve the supporting processes to deliver business value.

- **Security:** This pillar encompasses protecting your data and assets to improve your security posture.

- **Reliability**: This pillar provides the ability of a workload to execute the functionalities consistently 24/7. Reliable applications can also be operated and tested throughout their total life cycle.

- **Performance Efficiency**: This pillar allows your applications to use computing resources efficiently to meet the system requirements and maintain efficiency as demand changes.

- **Cost Optimization**: Your applications can run and deliver business value at the lowest price point with this pillar.

- **Sustainability**: Your applications can address the long-term environmental, economic, and societal impact of your business activities.

Self-healing and highly resilient

Self-healing systems are not about using complex logic but about understanding the scenarios in which your systems will fail and overcoming such problems to minimize the impact of failures. A self-healing system can uncover errors in its operations and automate simple recovery steps to operate in a better functioning state. Errors can happen at the hardware, network, or application level. Short-lived failures such as network connectivity issues or failed database connections, and big events, such as regional outages, are focused on during the design of such applications.

These applications have the capability to document errors or problems in an **exceptions log** for further analysis. Self-healing applications incorporate design attributes that can resolve such errors without manual intervention. The applications that are capable of self-healing can detect, log, monitor, and respond to failures gracefully.

Self-healing applications have the following behavioral attributes:

- **Failure identification**: Large-scale and distributed systems have a high probability of having unpredictable outcomes that can disrupt their functioning state. It is important to identify such areas before your systems run in production mode. **Chaos engineering** is an approach that can facilitate the experimentation of fault injection and uncover systemic weaknesses.

- **Failure prediction**: The underlying concept of this attribute is that you know your application in and out so that your teams can understand and dive into error scenarios. Automatically detecting or predicting failures will help you trigger healing mechanisms promptly to ensure the minimum level of functionality in your systems. Testing with fault injection and chaos engineering are well-adapted mechanisms to test the scenarios that could lead to failures.

- **Failure handling**: Building retry logic into your application to handle failures such as service timeouts and protecting dependent services from failing using circuit breakers when an operation is likely to fail are some of the recommended techniques to incorporate into your systems. In addition to the preceding capabilities, applications can continue serving end-user requests in an acceptable manner. Even when one or more of the dependent components fail to function, the overall system must be resilient to avoid any cascading failures.

Hyper automation

Modern applications enable you to move faster in terms of releasing more functionality. **DevOps** is a combination of cultural philosophy, automation practices, and tools that transforms the culture and mindset of the development and operation teams. Increasing efficiency through automation generates repeatable motion that speeds up the flywheel. Modern applications leveraging the DevOps model strategy can be more flexible and enable quicker innovation.

Automation in each step, as shown in the following diagram, can reduce the chances of delays and manual errors:

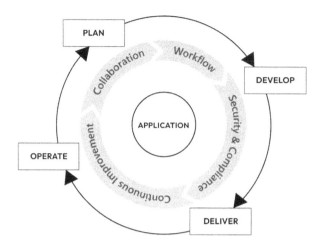

Figure 6.2 – The life cycle of modern applications

The development and operations teams are no longer operating in their own silos, but the two teams are merged and get to work across the life cycle of the application, right from development and testing to deployment and operations. The teams practice automating every step that has historically been error-prone and caused delays. The technology stack and tooling enable the engineers to accomplish tasks such as deploying code or provisioning infrastructure.

To move faster, the following practices can be implemented.

- **Continuous integration/continuous deployment (CI/CD)**: CI/CD processes enable automating the release pipelines to release high-quality code faster and more often

Automated release pipelines can accelerate the release process from code changes and build requests to testing and deploying.

The following diagram depicts how CI/CD delivery can help modern applications be deployed through an accelerated delivery pipeline:

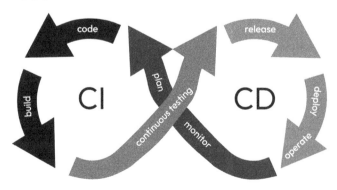

Figure 6.3 – CI/CD software practice

With CI, building, and testing, the code is automated fully so every time the developers commit the code, changes are validated and merged to the master branch of the repositories. In addition, packaging and building the artifacts is streamlined, helping the developers to get more time to make changes and contribute to improving the software.

With CD, each change that passes the automated tests is automatically placed in a desirable environment and deployed, resulting in faster production deployments. This is an extension of CI where the code changes are automatically deployed to a testing and/or staging environment after the build stage. With automated tools in place, you can decide to release updates to your applications daily, weekly, fortnightly, or to a schedule that best suits your business requirements.

CD automates all stages of your production pipeline, which goes all the way to your customers. With no human intervention, CD is an excellent means to accelerate pipeline releases unless a failed test is encountered. The following diagram shows how each of these practices relates to each other:

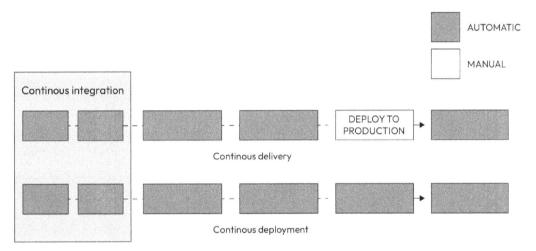

Figure 6.4 – Continuous integration, delivery, and deployment

In summary, the combined practices of CI with continuous delivery or CD allow the development teams to deliver code changes more frequently and reliably.

- **Infrastructure as code (IaC)**: IaC is the practice of codifying and managing the underlying IT infrastructure as software. The developers and operation teams can automatically manage, monitor, and provision resources with this approach. Historically, these steps caused a lot of delays, which involved the manual configuration of hardware devices and operating systems. Automating IT processes that are static in nature to make the deployment processes more versatile and have adaptive provisioning capabilities is the concept behind IaC.

The following predefined steps are fully automated using IaC:

I.   Provisioning resources

II.   Configuring instances

III.   Configuring and deploying workloads into the instances

IV.   Connecting the dependent services

V.   Monitoring, observing, and managing the deployments

Modern applications typically leverage the IaC tools that are suitable for them, enforce automation, and orchestrate the deployment of the preceding steps, which helps create infrastructure in a repeatable, reliable, and consistent manner.

In the next section, we will discuss the key steps for companies to transform and update their legacy systems.

Understanding key strategies for AppMod

The intent of transforming legacy applications with newer languages, frameworks, and infrastructure platforms is to provide improved user experiences and to meet the ever-growing demands of the digital market. Today, businesses of all sizes are expected to deliver software that is robust, scalable, secure, and accessible from any device at any time. Historically, the majority of the IT budget for any company goes toward managing and maintaining its legacy systems. When it comes to modernizing, driving a low-maintenance future for your IT portfolio is the key criterion for any enterprise.

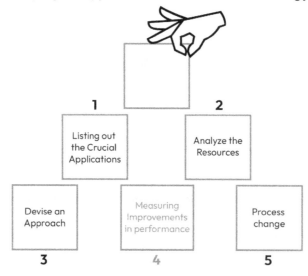

Figure 6.5 – Developing an AppMod strategy

To enable your teams to implement the modernization efforts effectively, you need a pragmatic modernization strategy that includes the following five elements, as shown in the preceding diagram.

Listing the critical applications

When your enterprises have multiple work streams, transforming and modernizing can be a long journey. Ensuring that your team's focus is on the critical applications that can impact your business and setting clear objectives allows them to execute their activities effectively. In addition, list such mission-critical applications and build your modernization roadmap early in the journey.

By doing so, you set objectives to help create performance benchmarks and ROI standards. You get to measure effectiveness periodically and fine-tune your strategy wherever needed to keep your roadmap on track.

Identifying the key resources

Once you have the modernization roadmap ready, identify the key people and assign them tasks based on their skill sets. Having **subject-matter experts (SMEs)** for each application in your roadmap will allow your team to have a sense of ownership and accountability. These SMEs can help coach new employees as well as provide feedback. As a thought leader, it is your responsibility to ensure you are assigning the most appropriate and fitting responsibilities to each of these SMEs. This will help you foster a support system and get quicker results. Separating the areas of responsibilities will also help you identify whether any gaps need to be addressed.

Designing an approach

By now, you know that there are many approaches (the six R's of migration to implement transformation). Some of the most common approaches are listed here:

- **Rewriting/refactoring**: This is where you can rewrite an original system and leverage software technologies that are better suited to your business.

- **Porting/rehosting**: This approach is where you simply move your workloads to a cloud platform and improve scalability and maintainability.

- **An incremental approach**: This approach combines the best of both the preceding approaches. Start with refreshing the platform, extend your current system to the cloud, and then refactor it to leverage modern technology, offering better experiences to your customers.

Each of the preceding strategies is suitable for a given use case, and analyzing the pros and cons of each approach will enable you, as a leader, to pick the right strategy.

Defining the success criteria

Defining key results and success criteria is important to be able to measure your transformation success. It gives you insights into your team's performance and where you stand on your journey. Measuring improvements with data-driven approaches will allow you to steer any necessary changes to the work streams and identify resources or knowledge gaps that may have come up during the journey.

The **key performance indicators** (**KPIs**) can be difficult to know and track. Treat your business outcomes as the real measurement and work backward to identify the parameters that are responsible for increasing the business value and reducing the operational risk/overhead.

Continually reassessing your approach

As we discussed earlier, modernization efforts can often be a long journey, so it is essential to regularly assess your organization's applications and see what cloud technologies and cloud-native architectures are the best fit. When you look at these emerging technologies, having a holistic approach is the best thing you can do for your company's business. Look at the greater value in terms of business benefits, reducing risk, and increasing efficiency when you incorporate modern technologies. Compare it with in-house versus cloud offerings to embrace the benefits of the infrastructure and application patterns.

In the next section, we will deep-dive into the concept of microservices and the various approaches that you can leverage to transform monolithic applications into microservices.

Breaking monolithic applications into microservices

Before we get into the topic of breaking a monolithic application into microservices, let us understand what a **monolithic architecture** is and the need to go through the process from a monolithic architecture to microservices architecture.

What is monolithic architecture?

Applications that are designed using *single-tiered* or *monolithic architectures* have all their functions in a single package to deploy and run. For example, when you develop a server-side Java application, it has different layers in the architecture, namely the following

- **Presentation layer**: This is where all the end user HTTP/HTTPS requests are handled, and responses are sent back

- **Business logic layer**: This is where the application's business logic is handled

- **Persistent/Database layer**: This is where the data access objects are persisted in the database

- **Integration layer**: This is where integration with other services is handled via messaging or a **representational state transfer** (**REST**) API

All the preceding layers are packaged and deployed as a single artifact in a monolithic architecture. This approach is simple to develop, test, deploy, and scale horizontally and has historically worked well for small-size applications. Monoliths are relatively simple and familiar, as only one executable program needs to be maintained:

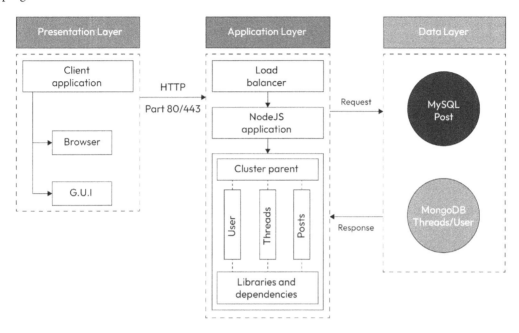

Figure 6.6 – A monolithic architecture

The preceding diagram shows a traditional way of building applications as a single and indivisible unit.

The benefits of monolithic architecture

Monolithic architecture offers several advantages when it comes to management and operational capabilities, including the following:

- **Simplicity**: Simple to develop for small-size projects
- **Easy to debug and test**: It is easy to implement end-to-end testing by using tools such as **Selenium UI** testing
- **Fewer cross-cutting concerns**: Aspects such as **logging**, **error handling**, and **caching** can be easily applied
- **Simple to deploy**: Only one packaged application needs to be copied to the server
- **Simple to scale horizontally**: It is possible to run multiple copies of the packaged application behind a load balancer

However, the shortcomings of monolithic architecture mainly vary with the complexity of the systems that you want to build for your business.

The challenges of monolithic architecture

Monolithic architectures may not be the right fit for all applications, and it is important to consider the following drawbacks before deciding:

- **Difficult to maintain complex systems**: For large distributed systems such as e-banking or online payment processing systems, it becomes challenging to make changes quickly and correctly to the code base using monolithic architecture

- **Long release cycles**: This is where the size of the application can slow down the startup time, and your teams end up redeploying the entire application on each update

- **Difficult to scale**: If your monolith has conflicting resource dependencies, it can be difficult to scale

- **Cascading failures**: This is where potentially one issue such as a memory leak can bring down the entire application

- **Difficult to adopt technologies and advanced approaches**: This is because the language or framework can be a constraint

- **Costly to refactor**: It can be costly to refactor applications with several modules

In short, consider building applications using a monolithic architecture for small, simple applications and have no business risks upon doing so. In the next section, we will look into the most popular architecture that can be a good choice for complex and distributed applications.

What is a microservices architecture?

With a microservices architecture, the application is broken down into smaller independent units where each unit is single-purpose. These units are all interconnected services and have their own architecture along with various connectors or adapters. Typically, the microservices would expose a REST, message-based API, or implement a web **UI**.

The following diagram shows how a microservices model is divided into smaller autonomous units:

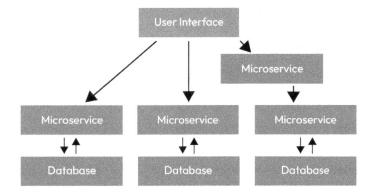

Figure 6.7 – Microservices architecture

Instead of sharing a single database, each microservice has connectivity with its own database. This is essential when you want to leverage the advantages of loose coupling, and each microservice can benefit from choosing the database type that best fits the business needs.

The importance of loose coupling in a microservices architecture

The idea behind microservices architecture is **loose coupling**. The main characteristic of loosely coupled systems is that the microservices should know little about each other. The following diagram depicts how tightly coupled systems that have several dependencies are different from loosely coupled systems:

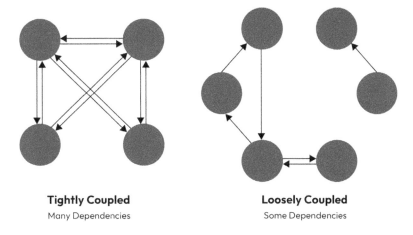

Figure 6.8 – Tightly coupled versus loosely coupled

A loosely coupled system has components that do not affect each other and has several benefits, as detailed here:

- **Ability to evolve faster**: Loosely coupled systems can make it easier for the development teams to make changes easily and push new solutions that can easily adapt to environmental changes.

- **Increased efficiency**: This is the ability to break or reconfigure the link between services, which helps increase the optimum efficiency of the overall architecture and also reduces the coordination costs.

- **Increased agility**: This is where development teams can iterate on a small, focused set of features and yield results quickly.

- **Ability to deploy faster**: This is where the microservices model removes dependencies and increases the frequency and stability of deployments.

Let us look at an example of a system that is built based on a microservices architecture.

An example of a microservice

Let us take an example of an online shopping system consisting of multiple modules. Let us look at the different components that it comprises and what each microservice is built for:

- **Authorization**: This service authorizes the customer to be able to purchase products

- **Order**: This service is responsible for processing the order when it receives one

- **Catalog**: This service is responsible for managing the products and keeping a check on the inventory

- **Cart**: This service manages the customer's cart and leverages the data source from the Catalog service

- **Payment**: This service is responsible for managing and authorizing payments

- **Shipping**: This service is responsible for shipping the successfully ordered items

The following diagram depicts how the online shopping service is designed using the microservices model:

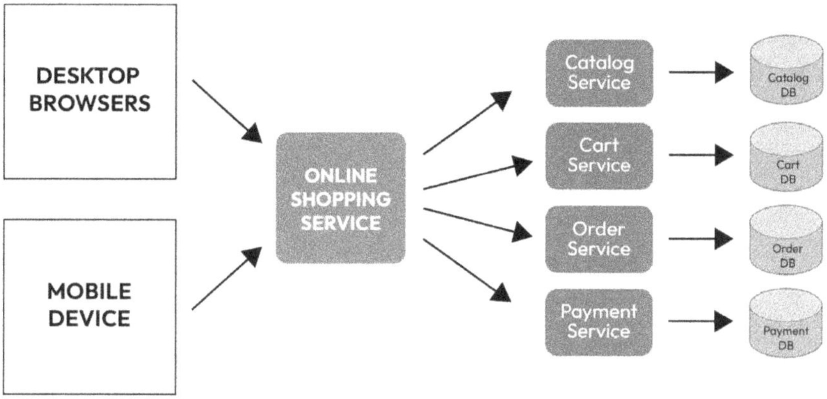

Figure 6.9 – Microservices architecture for an online shopping service application

In the next subsection, let us take a look at the advantages of creating large distributed systems using the microservices architecture.

The benefits of a microservices architecture

There are several benefits of using microservices architecture, out of which the following are the most noteworthy:

- **Increased scalability**: Microservices can be deployed and updated independently, which gives more flexibility in adding, removing, updating, and scaling individual microservices. When scaling happens to one service, the other services remain undisrupted and do not compromise the application. By scaling the specific services that need it, businesses can save a lot of costs on resources.

- **Improved fault isolation and resiliency**: The microservices in an application are less likely to be negatively impacted when one service is experiencing failures. Many large distributed microservices architectures tend to have dependencies, so it is a best practice to protect the application from a dependency failure leading to a shutdown.

- **Ability to build technology-agnostic applications**: Developers have the flexibility to build each microservice in any programming language and use polyglot databases. Any platform can be used, thereby increasing flexibility and encouraging different skill sets on the team.

- **Improved data security and compliance**: Microservices encourage the use of APIs, which are secure and ensure sensitive data is protected. A secure API ensures that the data is processed and accessible to only authorized applications, users, or servers.

If the microservices are responsible for managing healthcare- or finance-related data or any other kind of sensitive and confidential data, a secure API provides developers with an easy way to achieve compliance, such as with the **Health Insurance Portability and Accountability Act (HIPAA)**, **General Data Protection Regulation (GDPR)**, or any other data security standard.

- **Faster time to market**: The pluggable nature of the microservices makes it easier and faster for application development and upgrades to existing features. You can quickly build or change a specific microservice, and plug it into the architecture with the least risk, coding conflicts, or service outages.

  You also get increased support for CI and CD and build a future-proof product that can be iteratively evolved.

- **Increased business agility**: This is the ability to experiment with new functionalities with a microservice-based architecture, which gives you greater agility to respond to new business demands and deliver focused pieces of functionality quickly.

- **Two-pizza development teams**: The microservices architectural style supports the famous *two-pizza* development team structure, originally pioneered by Amazon. This helps in assigning ownership to a small team; individual teams get to work more closely, stay focused, and be accountable for the outcomes. This helps in improving work efficiency and producing higher-quality products.

While there are tons of benefits of microservices, for some use cases, building applications with a microservices framework can introduce increased complexity and effort. Let us take a look at some of the challenges of microservice-based architectures in the next subsection.

The challenges of a microservices architecture

Here are some of the common challenges that businesses experience while adopting microservices:

- **Increased complexity**: The numerous services involved in microservices increase the difficulties in managing them. An inter-process communication mechanism needs to be devised to ensure that you have proper communication between microservices when handling requests back and forth. Handling dependency management across different services effectively is recommended to avoid cyclic dependency.

- **Achieving data consistency**: Each microservice maintains and manages its data, giving rise to the chances of data duplication across multiple services. Because of this autonomous handling of data, redundancy issues can proliferate. Any alteration made to the schema can lead to changes in other dependent microservices because of several independent data storage solutions that exist. This can cause a challenge in managing data consistency across microservices.

- **Cross-cutting concerns**: Ensuring that externalized configuration components such as logging, monitoring, metrics, and health checks are implemented correctly can become a hassle if not implemented properly.

- **Testing**: Microservices-based applications make it much harder to test independently deployable components. The testing strategies need to be planned to suit the microservices model, and this requires a thorough understanding of how the architecture works.

You may be asking yourself which architecture to go with. The next subsection will throw some light on when to choose monolithic versus microservices architecture.

When to use microservices?

To answer this question, let us take a look at the scenarios when you want to choose a monolith versus scenarios that are suitable for microservices architecture.

The following diagram gives a holistic overview of how monolithic architecture is fundamentally different from microservice-based architecture:

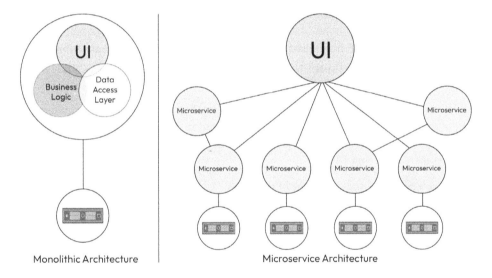

Figure 6.10 – Monolithic architecture versus microservices architecture

Here are the top three reasons why you should go with microservices:

Need for zero downtime

If your business strives for a platform where mission-critical applications can take in new functionality without changes to the rest of the system and any downtime, then microservices allow you to do so. The need for zero downtime is a vital aspect for **Software-as-a-Service (SaaS)**-based businesses where their applications or products are up and running 24/7.

Need for data isolation

Industries such as healthcare and finance require their products to be **Payment Card Industry** (**PCI**) and **personally identifiable information** (**PII**)-compliant. Specific data security and compliance standards also need to be adhered to. It is easier to isolate the data in the microservices model, allowing you to achieve data processing separately. In addition, it is easier to provide greater oversight, governance, and auditing for the separated-out data, thereby making it favorable for running applications that have data regulator constraints.

Need for autonomy

Collaboration, communication, and continuous learning are key aspects that cannot be overlooked for effective software development. With the microservices model, responsibility can be distributed across teams where they get a sense of ownership and develop products autonomously.

Microservices driving powerful and optimized technical services is made possible through a two-pizza team model where the *you build it, you own it* mindset is fostered and encouraged. Teams become responsible for the development, from building to releasing, allowing them to make optimal choices to speed up release cycles. This is historically difficult with monolithic architectures, given the exponentially large teams.

Although microservices are sweeping digital businesses and have become instrumental in modernization, it is important to analyze the practicality of the implementation before fully going down the path of developing microservices-based systems. In addition, it is important to weigh the costs and the benefits to ensure that you are achieving a ROI.

Now that we have a complete understanding of monolithic and microservices-based architectures and their benefits and challenges, let us take a look at the next subsection, which talks in detail about how to transform and evolve a monolith-based system into microservices-based architecture.

Splitting the monolith into microservices

If you are reading this section, you are embarking on the journey of the microservices ecosystem. Decomposing a monolithic application and transforming it into a microservices architecture while maintaining business as usual is a worthwhile journey but not an easy one. We have learned in the prior sections that if you aspire to run applications that meet aspects such as increasing the scale of operations or escaping the high costs of maintenance, an ecosystem of microservices is the recommended approach.

Some of the common challenges that lead architects face are deciding what capabilities to decouple and when and how to incrementally transition the applications. Before starting down the microservices path, ensure your teams are prepared for a certain level of operational readiness. Make sure that you have the groundwork ready in terms of the following:

- CD pipelines in place to build, test, and deploy services

- The ability to debug and monitor distributed architecture

- A dedicated infrastructure layer such as a service mesh to run a fast, reliable, and secure network of microservices

- A chosen container orchestrator (which we will cover in-depth in the next chapter)

The architectural challenges of detangling a monolith can be addressed through some of the proven techniques that I will share here. These techniques can guide the teams, including developers, architects, and technical leads, to make optimal decomposition decisions.

The main driver for decomposing a monolith is to be able to release the features or bug fixes in an agile manner and manage them independently. As we go through the next set of guidelines, it is necessary for you as a leader to closely evaluate whether each of the following recommendations is the right path for your existing monoliths.

Start with a simple capability to decompose

Start with features and functionalities that are decoupled from the monolithic application and don't require changes to the application that is currently using the monolith. Another recommendation is not to start with a service that needs a data store. Make sure your teams can build out the minimal infrastructure needed to deploy this service as an independently deployable unit. Simple edge services are a good candidate to start with.

In the online shopping service example, the Authorization service handles authorization and authentication to allow end users to access the system. At the beginning of the journey, it is good to use services where the operational prerequisites are easy to practice. Once your teams are used to these approaches, they can address the key aspect of splitting the monolith.

Decouple dependencies

One of the fundamental principles is that teams need to look to remove monolithic dependencies within the newly created microservices. That will help in producing a fast and independent release cycle. Dependencies such as data, logic, and APIs within a monolith can add to the complexity and result in high costs and a slow pace of change capabilities.

The order in which developers want to tackle decoupling is also a vital consideration. You cannot always avoid dependencies, but picking a service where a new API can be exposed from the monolith is important.

If we have to pick the next service to decompose from the online shopping service monolith, it will be the Catalog service because it has the least number of dependencies in terms of the order or payment service.

Decouple sticky features

By this stage, the development and operations teams are building microservices and are at the point where the remaining sticky services can be addressed. The main idea of this guideline is to identify a capability within the monolith that is causing the stickiness; it is sticky because of a not-so-well-defined concept. Next, deconstruct such capabilities into well-defined modules and then build them into separate services to make progress in these scenarios.

Incrementally extracting such modules from the stickiness of the monolith and building one service at a time can yield the best results. As an example, the Payment module can be extracted into a new service and then refactored, followed by the Orders microservice, and this is then repeated.

Decouple the data

The data store is a tightly integrated layer with brittle dependencies that needs multiple modules dependent on it to be released together. Imagine the online shopping service and its functionalities depending on one single backend system centrally storing all the data and holding its state. Data remains locked in one schema, and the storage system and its dependencies remain untouched as a result. Simply put, you can only go that far with architecting microservices without decoupling the data.

To address this issue, move out the capabilities vertically and, in essence, decouple the core functionalities from their data and separate all the user-facing frontend applications into new microservices exposing APIs. By doing so, multiple applications will read and write data into the decentralized data stores. Incorporate a data migration strategy that considers aspects such as data synchronization during migration, logical versus physical schema decomposition, latency, and so on.

Decentralized data management is the core of building microservices, and **polyglot persistence** is an approach that is most frequently used with microservices. This approach drives the usage of multiple data storage technologies that best fit individual applications or components. Picking the right data storage type for the right use case is the idea behind polyglot persistence.

Now that we have discussed the common techniques, let us move to the next subsection, which discusses a widely used pattern for decomposing a monolith system.

The Strangler pattern

One of the most commonly used approaches for modernizing a legacy monolith application into a microservices architecture is the **Strangler pattern**. Let us dive deeper into this approach and understand the technique in detail.

The idea behind the Strangler pattern is incrementally transforming a legacy system by separating out specific pieces of functionality and replacing them with new services. The older components that are replaced will eventually be strangled from the monolith and decommissioned. The following are the three main steps in this pattern:

- **Transform**: In this step, a new service is created that incorporates modern tools, technologies, and platforms.

- **Run in parallel**: The existing functionality is left running, and some of the traffic is redirected to the new service. This step is incrementally iterated until all of the traffic is redirected to the new functionality.

- **Replace**: In this step, the old functionality from the existing monolith is removed and decommissioned.

This popular design pattern advocates incremental steps versus *everything at once*, which is more error-prone and complex. You can always change the direction in the second step, which gives the development teams more room for experimentation. This pattern lets you incorporate test-driven development to achieve high-quality code. The next question you may have is understanding how to select the components that can be strangled first. Let's discuss that in the next subsection.

How to select the components to strangle first?

If you are new to the strangler pattern, this section will shed some light on the guidelines that you can incorporate as best practices and that can increase the chances of executing the strangler pattern successfully:

- Play it safe and select a simple component that will ensure you get the experience and practical knowledge to get comfortable with the challenges and best practices before handling a complex component.

- Pick a component that has good test coverage and less technical debt, as that will give the development teams confidence in moving ahead with the process.

- Pick a component that is suited for scalability and the cloud.

- Pick a component that frequently has business requirements coming in, which means preparing it to be agile and deployed a lot more regularly. By strangling such a component, you don't have to redeploy the entire monolith along with it.

Overall, you need to focus on the high-level goal of reducing the application's overall complexity and increasing its agility to deliver the business features faster.

The modernization journey is not an easy one, and the Strangler pattern assists you in making decisions that are risk-free since you will be dealing with one component at a time incrementally. In addition, automated CI/CD enables you to deliver business capabilities faster and makes it easier to deploy these microservices.

In the next section, we will look into some of the most common microservices architecture design patterns intended to help you design and implement microservices using a combination of these patterns that best suit your needs.

Microservices architecture patterns

Microservices architecture is a trending choice for many companies incorporating modern application development practices. Microservices patterns that are designed with different specifics and priorities can come with their own challenges. Although your requirements drive the decision of which pattern is the best fit, you will be able to achieve your microservices architecture goals of hitting the targets of speed, scalability, and cohesion if it is implemented correctly. Understanding the design patterns will empower you to tackle the orchestration challenges. Before we get to some of the common design patterns, let us look at what a design pattern is.

What is a design pattern?

A design pattern is the structure of a series of recommendations to solve a certain problem. The structure lets you define the issue, what difficulties you had in the past, and what the solution looks like. A design pattern helps you solve many complex issues commonly encountered in microservices design. This sets the context of previous experience and routines to help you solve specific problems faced in the design more efficiently. Each pattern is better at solving a particular problem than others; this knowledge will give you a holistic understanding and help you pick the right pattern for your use case.

Microservices design patterns allow development teams to accelerate the development phase and can be categorized as shown in the following diagram:

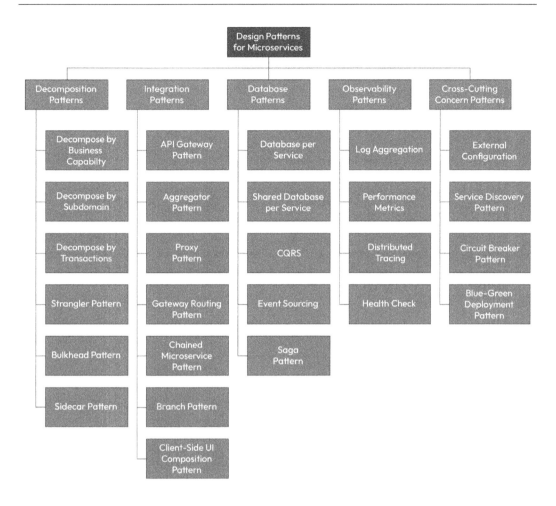

Figure 6.11 – Microservices design patterns

Let us do a detailed breakdown of the decomposition design patterns and understand how these patterns can be used while developing microservices.

Decomposition design patterns

This pattern provides key principles for decomposing applications into smaller microservices and how to do so incrementally and logically. Here are some of the most popular patterns:

- **Decompose by Business Capability**: By defining a microservice corresponding to a particular business capability, architects can categorize application components into multi-level hierarchical structures. That way, the architecture is highly stable, and development teams can work independently and be organized around functional features instead of technical features.

The caveat to this pattern is that the business capabilities need to be well understood, and the domain objects need to be defined corresponding to the business capabilities, which can add overhead to the planning.

- **Decompose by Subdomain**: The microservices are defined by **domain-driven design** (DDD) subdomains based on the business functions. Each domain can have multiple subdomains referring to different line-of-business functions or different areas within a single business. The decomposition of a monolith using this pattern is usually as stable as the business subdomains. Development teams can work independently, and cross-functional teams are aligned around functional features instead of technical features.

- **Decompose by Strangler**: The Strangler pattern is an approach to decomposing monolith applications incrementally into microservices. This pattern was originally invented by Martin Fowler and has gained traction ever since microservices became popular. When we investigate application architectures, they are a mix of modular and non-modular ones. The non-modular ones are hard to navigate, and the strangler pattern can be used as a guide to navigate the complexities of non-modular architectures. The Strangler pattern forces you to first create the modules in the monolith application and then spawn them into microservices for each specific module. The old component is decommissioned altogether and replaced by the new component to achieve high-quality code and efficiency.

The recommended approach is to identify simple components that are easy to refactor or modules that receive less traffic as compared to other modules. This two-step approach works well and is a safe approach for services that require refactoring and qualify for development time. Also having a rollback plan handy in case something goes wrong while refactoring is a must.

Decomposing the monolith using the previously mentioned patterns has its own advantages and disadvantages. Choosing the right pattern for a given application depends on factors such as your requirements and any dependencies of the system.

Fine-grained service-oriented architecture (SOA)

One of the most commonly used patterns used to minimize friction with **service-oriented architecture** (**SOA**) divides the design into smaller, fine-grained segments aligned to work on the same principle as the SOA. If you are experiencing issues such as inefficient inter-process communication, management, governance, or monitoring issues, fine-grained SOA will enable you to tackle the complexity involved in managing cloud-based applications.

Use case

This pattern can be used when you have a massively complex system to work on where minute details need to be managed efficiently. The complex system is divided into numerous individual units, and each of those units has a better and faster management system to control it easily.

Best practices for modern application development

Building modern applications has enabled a shift in focus toward architectural patterns, operational models, and software delivery processes. While these shifts are transformative, the journey doesn't have to be brutal. You will want to consider the best practices as you build modern apps. These best practices have been captured from interviewing various leaders that have enabled their organizations' modernization journey, as well as from the experience of building applications on AWS. You can perceive these best practices as a starting point and we will cover them in the following sections.

Enable accountability and innovation

Enabling innovation to deliver better customer outcomes is the starting point for modern application development. When developers are building products and responsible for running and maintaining them, you are making them accountable for the health of the whole product and not just pieces of it. This gives a sense of ownership and autonomy to the teams. This ownership of the complete application life cycle includes taking customer input, planning the roadmap, and developing and operating the application. Your development teams are empowered to deliver new features, thereby creating a flywheel of motivation and risk-taking culture in an environment of confidence and trust.

Build microservices wherever suitable

As a leader, you are headed in the right direction by building a strong culture of ownership, but your development teams will struggle to scale up if your applications are not built to handle dependencies and cannot adapt to the growing scale. Building microservices architectures for applications enables you to divide complex applications into components that a single team can own and run independently. As a leader of a fast-growing business, you need to ensure your company is innovating, and moving to microservices can enable you to reduce the overhead and move quickly to production.

Automate wherever possible

Your teams are enabled to move faster when the release pipeline is not bogged down and driven manually. Look for opportunities to automate and remove delays and manual errors. CI/CD has become the foundational floor for building modern applications and can help you achieve millions of deployments and grow faster every year. Teams that incorporate CI/CD have shipped more code, and it is historically proven that the commit-to-deploy ratio is several hundred times faster and more frequent.

Additionally, application security should be treated as the highest priority, where security and compliance should be included in every stage of the application development life cycle. In general, the microservices architecture patterns can improve security by providing fine-grained control over the security policies of each microservice.

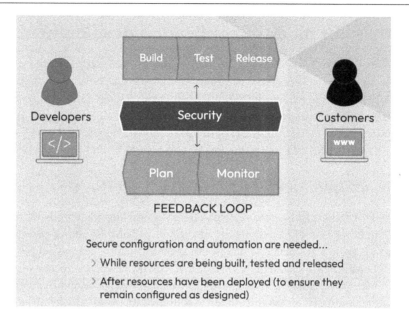

Figure 6.12 – Securing the entire application life cycle by automating security

However, there is also the risk of the increased surface area that you need to protect. Moving deeper into the software engineering space and addressing security in each step of the life cycle can yield better results without slowing down the application cycle. Automate the process of testing those security features to ensure they remain configured as designed.

Use IaC

If your application architecture is modular and releases features quickly, there are a lot of moving parts. Manually making changes to the environment can easily introduce errors, which can break your application's state. And as your application scales, the ability to scale the infrastructure seamlessly is challenged. The key is to keep your resources running as expected while you maintain a consistent environment that can be easily scaled or modified.

IaC, as described in the previous sections of this chapter, allows you to provision your resources safely and predictably, thereby eliminating the preceding challenges.

Add observability

The idea of **observability** is becoming increasingly important in a microservices landscape. Observability aims to provide insights into the behavior of the microservices and is particularly important to see how inter-process communication is happening. A primary microservices challenge is understanding how individual services interact. Observability is particularly critical when you need a way to see when communications fail, and how services interact with each other at runtime needs to be monitored.

Observability is often conceived as monitoring and is used interchangeably, which is not correct. Monitoring provides a way to capture the health of your system, whereas observability takes those findings and provides data to know what to ask next. In a microservices world, observability gives developers, engineers, and architects various tools to observe the ways these services interact. This is usually carried out during specific phases of development or across the entire life cycle of the application. With complex and large architectures, this can become particularly challenging. Observability over production-grade applications enables smooth operations and drives innovation with continuous code releases.

The AWS landscape for AppMod

AWS offers a platform to run modern applications that can balance managing technology and building new capabilities. Modern applications that are built with modular architectural patterns, serverless operational models, and agile development processes allow you to innovate faster and accelerate your time to market. Modernization in AWS is broadly classified into the following categories, where we will highlight some of the most prominent AWS offerings. In the next chapter, we will deep-dive into each compute offering that AWS has.

Serverless

Serverless offers technologies for running code, managing data, and integrating applications, all without having to manage the underlying infrastructure or servers. With automatic scaling, built-in high availability, and a pay-per-use billing model, these technologies allow you to focus on your business instead of management tasks such as capacity provisioning, patching, and so on.

AWS Lambda

AWS Lambda is a serverless and event-driven compute service to run code for any type of application or backend service without the need to worry about managing or instantiating the underlying servers. You can trigger Lambda from over 200 AWS services and SaaS applications and only pay for what you use. Some of the features of AWS Lambda include the following:

- Bring your own code

- Package and deploy functions as container images

- Connect to relational databases

- Connect to shared filesystems

- Orchestrate multiple functions

- An integrated security model

We will discuss the serverless service offerings of AWS Lambda in *Chapter 7, Application Modernization – Compute*.

AWS Fargate

AWS Fargate is a serverless, pay-as-you-go compute engine to build applications without managing servers. AWS Fargate is compatible with Amazon **Elastic Container Service** (**ECS**) and Amazon **Elastic Kubernetes Service** (**EKS**), which we will cover in depth in the next chapter. Some of the features of AWS Fargate include the following:

- Web apps, APIs, and microservices
- Run and scale container workloads
- Support **artificial intelligence** (**AI**) and **machine learning** (**ML**) training applications
- Optimize costs

We will review AWS Fargate in *Chapter 7*, *Application Modernization – Compute*, in the *Serverless* section.

Application integration

Application integration on AWS is a suite of services suitable for enabling communication between decoupled architectures such as microservices, distributed systems, and serverless applications. By using these services, you get the benefits of decoupled architecture and an improved time to market.

Amazon EventBridge

Amazon EventBridge is a serverless event bus used to deliver real-time data and connect applications to multiple data sources. EventBridge offers the flexibility to configure routing rules and build event-driven architectures.

> **Event-driven architecture**
>
> Referred to as *asynchronous* communication, event-driven architecture is a software design pattern that gives applications the ability to detect key business transactions, *events*, and process them in (near) real time.

EventBridge consists of building blocks such as events, event sources, event buses, rules, and targets.

An event is a real-time change that affects the application, its data, or the environment. An event source is the source that triggers the event. An event bus is a construct that receives the event from various event sources. A rule is a JSON snippet that controls the routing of the events.

The target could be one of several AWS services such as EC2 instances, Lambda functions, ECS tasks, **Simple Notification Service** (**SNS**) topics, and Amazon **Simple Queue Service** (**SQS**).

EventBridge works according to a publish/subscribe model where SaaS applications can publish events to the event bus and they are processed as events to one or multiple AWS target services.

EventBridge comes with a wide variety of support features, including the following:

- **API destinations**: These support many targets and integrations

- **Schema discovery and registry**: This centralizes and shares the event structure using popular programming languages such as Java and Python

- **SaaS integration with third-party consumers**: These include Datadog and PagerDuty

- **Event filtering and content-based filtering**: This is based on numeric comparison, prefix matching, and IP address matching

- **Automatic response**: It adjust to operational changes in AWS services

From a developer perspective, EventBridge has addressed integration complexities in the cloud world for SaaS platforms and identity providers, which are key platforms to integrate with.

AWS Step Functions

AWS Step Functions is a low-code, visual workflow service that developers can use to build distributed applications. Developers can easily automate IT and business processes using AWS Step Functions. Building data and ML pipelines can be streamlined using this service. With its native capabilities, developers don't have to worry about managing failures, retries, parallelization, service integrations, and observability. Instead, they can focus on higher-value business logic. Some of the features of AWS Step Functions include the following:

- Workflow configuration

- Service integrations

- Workflow abstraction

- State management

- Error handling

Step Functions work well with both synchronous and asynchronous business processes. As an example of an asynchronous process, developers can leverage this service as a serverless orchestrator between multiple services, such as AWS Lambda and Amazon SNS components, to process a credit card limit increase for a banking system.

As an example of a synchronous process, data processing systems can leverage Step Functions to configure and define a series of steps.

One of the most common use cases for Step Functions is that developers can leverage this serverless orchestration service to manage their workflows easily and focus on high-value business logic. This service works well for both synchronous and asynchronous business functions. An asynchronous use case could be the approval of a credit card limit increase where Step Function can act as a workflow orchestrator between multiple microservice components.

Amazon SQS

Amazon SQS is a fully managed message queuing service used by developers and IT architects to store, capture, and push messages asynchronously. This helps in decoupling and scaling microservices, distributed systems, and serverless applications.

SQS addresses the challenges and operational overhead associated with managing message-oriented middleware. SQS can be used to safely enable messaging between multiple parties and software components. Some of its features include the following:

- Standard queues and **first-in-first-out** (**FIFO**) to retain the messages in the same order that they were originally sent and received
- FIFO queues for high throughput and for maintaining the order of operations for critical events
- Unlimited queues and messages without requiring additional configuration or setup for enterprise-grade distributed systems
- Long polling and short polling to support immediate message processing, as well as polling multiple queues in a single thread
- Queue sharing to securely share SQS queues within AWS accounts
- Message locking to avoid duplicate message processing

Many enterprises find it useful to take advantage of SQS's decoupling features for heavy-duty batch-oriented systems for scale and reliability.

Amazon SNS

Amazon SNS is a fully managed messaging service for two types of communication: **application-to-application** (**A2A**) and **application-to-person** (**A2P**).

The A2A **publish-subscribe** (**pub/sub**) functionality is favorable for high-throughput, push-based, many-to-many messaging. This is suitable for distributed systems, microservices, and event-driven serverless applications. Your publisher systems can fan out messages to a large number of subscriber systems using SNS. This includes Amazon SQS queues, AWS Lambda functions, HTTPS endpoints, and Amazon Kinesis Data Firehose for parallel processing. The A2P functionality enables you to send messages to users at scale via SMS, mobile push, and email. Some of the features include the following:

- An event-driven computing model
- Message publishing and batching
- Message filtering
- Message fan-out and delivery

- Message durability

- Message encryption

SNS can be mainly used for event notifications and updating key stakeholders about time-sensitive information depending on what is critical to the business. Developers can rely on SNS to relay events within distributed computing applications.

Amazon API Gateway

Amazon API Gateway is a fully managed, elastic self-service with which developers can pay by use to dynamically create, publish, maintain, and manage APIs on the cloud. Developers do not have to set up EC2 instances and configure or install gateway software. They can directly use API Gateway and everything is handled in the serverless backend infrastructure. You can create RESTful and WebSocket APIs within polyglot frameworks.

API Gateway handles API life cycle management effectively and provides unique advantages that many developers and DevOps teams can leverage in the context of deploying microservices on AWS. Some of the features include the following:

- Support for RESTful APIs and WebSocket APIs

- Easy API creation and deployment

- SDK generation for Android, JavaScript, and iOS

- API life cycle management supports developers with dedicated endpoints for testing, development, staging, and production environments

Amazon API Gateway can be ideal for enterprises that are utilizing Amazon's **application lifecycle management** (**ALM**) services such as Amazon CodeCommit, CodeDeploy, and CodePipeline. If your applications are heavily running on AWS Lambda, then API Gateway makes integrating very easy.

AWS AppSync

AWS AppSync is a fully managed service that allows you to build GraphQL APIs and handle the heavy lifting of securely connecting to data sources such as AWS **DynamoDB**, Lambda, and others. Adding caches to improve performance, subscriptions to support real-time updates, and client-side data stores that keep offline clients in sync is just as easy. **GraphQL** helps organizations to develop applications faster and allows frontend developers to query multiple databases, microservices, and APIs with a single GraphQL endpoint.

AppSync can automatically scale your API execution engine up and down to meet API request volumes. Some of the features include the following:

- Real-time data access and updates

- Offline data synchronization

- Data querying, filtering, and searching
- Caching

Custom domain names, applications, and services that can benefit from a GraphQL API can be a good fit for AppSync where complex web applications can improve their load times. Saving time from setting up a **do-it-yourself** (**DIY**) gateway and still getting industry-standard best practices using AppSync to connect to the backend services makes it an ideal choice for AppSync.

Amazon Simple Storage Service (S3)

Amazon **Simple Storage Service** (**S3**) is an object storage service that offers industry-leading scalability, data availability, security, and performance. You can store and protect any amount of data for a wide variety of use cases using Amazon S3. S3 is suitable for building data lakes, websites, cloud-native applications, backups, archives, Machine Learning (ML), and analytics. Amazon S3 is designed for 99.999999999% (11 9's) of durability and stores data for millions of customers all around the world. Some of its features include the following:

- Storage management and monitoring
- Storage analytics and insights
- Storage classes
- Access management and security
- Data processing
- Data transfer

S3's data storage infrastructure makes it ideal for any enterprise to save costs while operating their web-scale computing systems and take advantage of the service's highly scalable, fast, reliable characteristics.

Amazon DynamoDB

Amazon DynamoDB is a fully managed, serverless, key-value NoSQL database ideal for running high-performance applications at any scale. You get built-in security and can automate multi-Region replication. You also get features such as continuous backups, automated multi-Region replication, in-memory caching, and data export tools. Some of the additional features include the following:

- **Atomicity, consistency, isolation, and durability** (**ACID**) transactions
- Encryption at rest
- Point-in-time recovery
- On-demand backup and restore

It is important to evaluate and choose the right type of database that you require for your enterprises that depend on high-scale database operations. DynamoDB is particularly useful for enterprises that expect high-performance reads and writes. Additionally, if you are operating mission-critical applications that require a high level of data durability and the least manual intervention to ensure your business functions are always available, then DynamoDB is right for your use case.

AWS targets relieving the most common pains for enterprises and database freedom is one of the aspects that AWS aims to offer.

Amazon Relational Database Service (RDS) Proxy

Amazon **RDS Proxy** is a fully managed database proxy that provides high availability when you are using Amazon RDS to build applications that have resilient and secure databases.

RDS Proxy is useful for applications built on modern architectures that have a large number of connections that need to be managed on their database servers. This can exhaust the database memory and compute resources, whereas Amazon RDS Proxy allows applications to pool shared connections and improve database efficiency and application scalability. The failover times for Aurora and RDS databases are reduced by up to 66%, and database credentials, authentication, and access can be managed through integration with **AWS Secrets Manager** and **AWS Identity and Access Management** (**IAM**).

It is simple to enable Amazon RDS Proxy for most applications and requires no code changes. You don't need to provision or manage any additional infrastructure, and the pricing is simple and predictable. You pay per **virtual centralized processing unit** (**vCPU**) of the database instance for which the proxy is enabled. Amazon RDS Proxy can be used for AWS database offerings such as Aurora MySQL, Aurora PostgreSQL, RDS MySQL, and RDS PostgreSQL.

Amazon Aurora Serverless

Amazon Aurora Serverless is a managed configuration offering for Amazon Aurora. Aurora is Amazon's MySQL and PostgreSQL relational database model built for the cloud. It delivers on-demand, auto-scaling, and highly available relational databases where all the maintenance, such as patching, backups, and replication, are handled for you.

Aurora Serverless is a good fit for applications that need to regularly scale and cater to unexpected traffic. Steep and unpredictable spikes in usage that would typically require manual compute and capacity adjustments are handled for you by Aurora.

In the next section, we will take a look at how a company was able to modernize its workloads using AWS services on the cloud.

Case study

ABC is a global leader in digital communications software, services, and devices for businesses of all sizes. ABC currently serves more than one million business customers and provides integrated communications solutions to companies of all sizes worldwide. ABC uses on-premises virtual machines to house its communication solution and encounters a variety of challenges, such as the following:

- High maintenance costs for on-premises infrastructure

- Unable to rapidly scale and handle customer demands quickly

- Low customer experience resulting in queuing of requests

- Lack of automation of customer onboarding tasks

To improve, ABC decided to migrate its IT communication solution to the cloud, on demand through an automated process. ABC used AWS CloudFormation to provision a new environment and migrate its service to the public cloud using AWS OpsWorks for Chef Automate. ABC refactored its applications to leverage a microservices API built with Python and Django to enable it to quickly spin up new environments on demand. The microservices API is deployed on Amazon ECS.

The company leveraged the following services from AWS:

- **Amazon ECS**: This is a container orchestration service for managing and scaling ABC's containers

- **Amazon Route 53**: This is a **domain name system** (**DNS**) web service for routing end users to ABC's applications

- **Amazon S3**: This is an object-based storage service for storing and retrieving any amount of data from ABC's environment

- **AWS CloudFormation**: This provisions all the resources needed for ABC's applications in an automated and secure manner

- **Amazon ElastiCache**: This provides in-memory data storage and sub-millisecond latency to power ABC's applications

- **Amazon Redis**: This is a scalable in-memory data store built to power real-time applications

- **Amazon RDS**: This is a persistence store for real-time transactions with ABC's end users

The detailed architecture is shown in the following diagram:

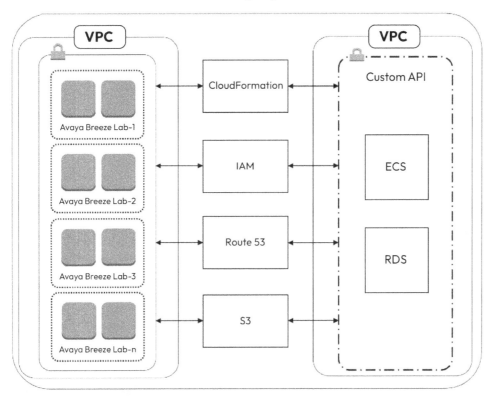

Figure 6.13 – Modernized solution architecture for ABC

The overall modernization helped ABC to reap the following business benefits:

- **Faster provisioning**: This enabled automatic provisioning of the new application across AWS regions with a simple API call

- **Scalability**: This provided the ability to scale up or down based on predefined activities or new feature launches

- **Improved operational performance**: This automation of infrastructure meant no human intervention was required for repetitive build management or resource onboarding tasks

- **Improved customer experience**: All the preceding changes helped ABC to double its customer base without adding extra resources to its IT teams, and with increased automation, it was able to cater to a broader set of end users in the community

This concludes the case study of a company's end-to-end IT portfolio modernization journey and the chapter.

Summary

In this chapter, we introduced you to AppMod, the key strategies involved, monolithic versus microservices architecture, design patterns, and best practices. The understanding you gained from this chapter will be especially useful as we dive into the next chapter about AWS's compute offerings, how to effectively leverage them, and when.

References

Martin Fowler's series on software development: `https://martinfowler.com/`

7

Application Modernization – Compute

The extraordinary pace of innovation opens up new opportunities to disrupt and maximize agility through application modernization. We also learned in the previous chapter that AppMod is a primary enabler for getting the most out of the cloud. You can bring your people, processes, and technology together into one streamlined platform while reinventing to scale up the ideas that will drive the growth for tomorrow. The ultimate objective of app modernization is for enterprises to shift away from "keep the lights on" operations and free up their budgets and teams to redefine their business value.

This chapter will focus on you becoming familiarized with the AWS compute platform and the offerings that will enable your application modernization efforts.

In this chapter, we will cover the following key topics:

- An overview of AWS compute services
- A case study for Amazon **Elastic Compute Cloud** (**EC2**)
- Diving deep into containers
- A case study for containers
- A case study for serverless

An overview of AWS compute services

By now, you are familiar with the concepts of cloud computing and the various cloud service models. Throughout this chapter and the next few, you will find details about the services that are most relevant for cloud migration and modernization. My goal is to build a foundation of enterprise modernization through which you can create your own modernization plan that will help revitalize your workloads.

> **Compute**
>
> Compute, in modern computing such as cloud computing, is the processing power or resources required by any applications or workloads to run and perform their activities.

Compute resources are infrastructure elements, including hardware and software, that allow end users to allocate their resources for problem-solving and solution creation. Cloud providers typically offer predefined and prebuilt machine types that come with ready-to-go customizations and configurations with a range of **virtual CPUs (vCPUs)** and memory to start running the applications quickly. This means that as an end user, you get to tailor your infrastructure to best suit your applications.

Let us take a look at the compute service offerings that AWS provides for you to manage your infrastructure on the cloud:

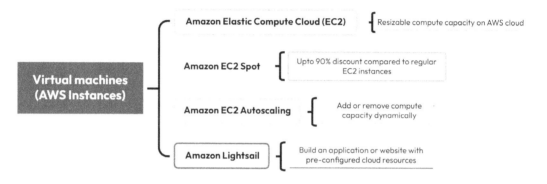

Figure 7.1 – AWS compute services – instances

Amazon Elastic Compute Cloud (EC2)

EC2 is a core compute web service that you can use to build and host your software applications in a secure manner. EC2 comes with flexible options where you can run your workloads on Linux, Windows, or macOS. Workloads of any type, be it legacy applications, websites, databases, storage, or physical or virtual servers that typically run in data centers, can be hosted on EC2.

Millions of enterprise applications have been migrated to EC2 since its inception in 2006, and the enterprises benefit from the global cloud presence that AWS offers. EC2 instances can be easily launched with pre-configured templates called **Amazon Machine Images (AMIs)**. AMIs come with a predefined software configuration template, such as an **operating system (OS)**, application server, and any applications that your workloads depend on. In addition to the security, high performance, reliability, and cost-effective infrastructure that the AWS ecosystem provides, EC2 is one of the most popular compute platforms that any enterprise can leverage and use as a first building block to migrate their workloads to.

For example, a leading open digital experience company is able to rely on the broadest and deepest functionality of Amazon EC2 by running their applications using the latest Amazon EC2 AMI and stand up their virtual servers within a few minutes, which otherwise takes several weeks, whenever the company needs to add compute and storage capacity to support their website for online marketing campaigns.

In order to fit the various needs and use cases of businesses, Amazon offers a wide selection of instance types. Let us take a look at what they are in the next subsection.

Instance types

Instance types are a wide collection of instances that come with a combination of aspects such as CPU, memory, storage, and networking capacity, allowing you to use the best instance that fits your use case.

There are five different categories of instance types:

- **General purpose**: The instances in this category offer a balance of compute, memory, and networking resources to cater to a diverse set of workloads. *M5* and *M5a* instances are well suited for small and midsize databases, backend servers for SAP, Microsoft SharePoint, and data processing tasks

  - *M5zn* is ideal for gaming applications or **high-performance computing** (HPC) that need extremely high single-threaded performance, throughput, and low-latency networking. *M6g* and *M6gd* instances are powered by AWS Graviton2 processors that are well suited for microservices, gaming servers, caching fleets, or midsize data stores

> **AWS Graviton Processor**
>
> Introduced in 2018, AWS Graviton is a new CPU developed by Amazon to deliver the best price-performance to run compute-intensive workloads. AWS Graviton3 processors are the latest in the AWS Graviton processor family as of 2022 and provide the best compute performance while supporting **Double Data Rate 5 (DDR5)** memory.

  Mac1 instances are powered by a combination of two components: Apple Mac mini, suitable for developing, building, and testing applications for Apple devices such as the iPhone, iPad, or Mac, and the AWS Nitro System for high-speed network bandwidth connection, suitable for fully integrated and managed compute instances with Amazon **Virtual Private Cloud** (**VPC**) networking.

  T2, T3, T3a, and *T4g* instances provide a baseline level of CPU performance, while *A1* instances offer significant cost savings that are suited for scale-out and Arm-based workloads.

- **Compute optimized**: This instance family is suitable for compute-bound applications that require high-performance processors. *C7g, C6g, C6gn, C6i, C6a, Hpc6a, C5, C5a, C5n,* and *C4* are examples of instances powered by AWS Graviton3 processors and the AWS Nitro System.

> **AWS Nitro System**
>
> In 2017, AWS designed this re-imagined virtualization with a combination of dedicated hardware and lightweight hypervisors. This is primarily to cater to the next generation of EC2 instances and enable customers to innovate faster and reduce costs. It also helps in accelerating your time to market and deliver improved functionalities.

Some use cases it is well suited to include batch processing workloads, media transcoding, high-performance web servers, **high-performance computing (HPC)**, scientific modeling, dedicated gaming servers, **machine learning (ML)** inference, and other compute-intensive applications.

- **Memory optimized**: The instances in this category are built to deliver fast performance suitable for workloads that have memory-intensive requirements. *R6g, R6i, R5, R5a, R5, R5n, R4, X2gd, X2idn, X2iedn, X2iezn, X1e, X1,* and *z1d* are some of the available memory-optimized instances that support enhanced networking, high memory bandwidth per vCPU, high networking speed, and memory encryption using Intel **Total Memory Encryption (TME)**.

  Instances that are launched using these instance types are ideal for running large enterprise databases and **electronic design automation (EDA)** such as static timing analysis and full chip gate-level simulation.

- **Accelerated computing**: These instances use hardware accelerators or co-processors to perform functions such as mathematical calculations, graphics processing, and data pattern matching. *P4, P3, P2, DL1, Trn1, Inf1, G5, G5g, G4dn, G4ad, G3, F1,* and *VT1* are some of the instance types that feature petabit-scale networking, low-latency storage, and intra-instance connectivity.

  These instances are suitable for ML training, computer vision, search recommendation, image and video analysis, forecasting, advanced text analytics, fraud detection, and many other recommendation engines.

- **Storage optimized**: The instances in this category come with the best price performance, always-on encryption, memory-optimized, and high network bandwidth, and support bare-metal instance size for workloads that benefit from direct access to the physical processors and memory. *Im4gn, Is4gen, I4i, I3, I3en, D2, D3, D3en,* and *H1* are the available set of instances that make it suitable for high, sequential read and write access to very large datasets on local storage.

Optimized to deliver tens of thousands of low-latency, random **input/output operations per second (IOPS)**, this instance type is best suited for running small to large-scale NoSQL databases, transactional databases, distributed file systems, data warehousing, analytics workloads, and distributed computing.

Now that you have learned about the wide selection of EC2 instance types available, let us take a look at how to leverage Amazon EC2 for cloud migration and modernization.

The migration process to AWS

AWS prescribes three phases to help with your migration process to the AWS cloud. Repeatability and predictability of the migration process are two core aspects that can be instrumental in accelerating the migration process of any organization. AWS focuses on providing automation and intelligent recommendations to simplify and accelerate each step of this three-phase migration process.

The following diagram depicts the three phases of how to migrate as you start your journey with AWS:

Figure 7.2 – Three-phase migration process to AWS

Assess

AWS provides three services to assess your organization's current readiness, identify the desired business outcomes, and develop the business case for migration.

With **AWS Migration Evaluator**, you can get a complete assessment of your on-premises resources and build a right-sized, optimized cost projection for running applications in AWS. This service provides you with details such as the **total cost of ownership** (**TCO**) on your actual utilization of resources on AWS. Here are some of the features of Migration Evaluator:

- *Inventory discovery* through a complementary agentless collector.

- *Insights* that provide visibility into the pre-migration assessment to the business and technical stakeholders, highlighting the estimated savings through rehosting on AWS.

- *Migration expertise* for additional assistance with your business case where a team of solution architects can help narrow down a subset of best-suited migration patterns.

- *Business case report* at the end of the migration assessment comprising five sections: assessment report, executive summary, on-premises costs, repurchasing scenarios, and next-step recommendations.

AWS Migration Hub provides the tools to migrate and modernize on AWS. It provides a single place to store IT asset inventory data required for discovery and planning. The status of the servers or databases that are marked for migration can be easily tracked, thereby helping to reduce any business risks. Here are some of the features of Migration Hub:

- *Import your on-premises server* details using the AWS Discovery Agent or AWS Agentless Discovery Collector

- *Build a migration plan* that provides you with a visual representation of servers and their dependencies

- Easily build *migration and modernization strategies* for your applications to transform at scale

- *Migration Hub Orchestrator* automates the migration process using templates and synchronizing tasks into a workflow to achieve your project goals

- *Migration dashboards* to show the latest status and metrics for your rehost and replatform migrations

With **AWS Prescriptive Guidance**, you get time-tested strategies, including guides and patterns to help with your cloud migration, modernization, and optimization efforts. The resources are developed by the technology experts and global community of AWS partners based on their years of successes and lessons learned through helping AWS customers achieve their business objectives on AWS.

Mobilize

This phase takes your migration plan to the next step where you get to address any gaps in the organization's readiness during the assess phase and refine your business case accordingly. We discussed in prior chapters and reviewed in detail the six migration strategies – relocate, rehost, replatform, refactor, repurchase, and retire. One critical aspect of developing your migration strategy is to collect your on-premises application portfolio data and rationalize it as per these migration strategies.

AWS offers several tools and services to help you make easier and better decisions as you plan your migration.

Migration Partner Solutions provides the required expertise, tools, and business/technology strategy to help enterprise organizations migrate their applications and legacy infrastructure to AWS. The following are the five software categories in which you can find technical enablement and functional content from the partner solutions:

- **Discover, planning, and recommendation**: So you can build a comprehensive migration plan using the six common migration strategies.

- **Business case analysis and TCO analysis**: To simulate migration and modernization approaches.

- **Application mobility**: To execute application migration and modernization to AWS.

- **Data mobility**: To transfer datasets and databases into target AWS resources.

- **Application monitoring and orchestration**: To gain insights into your application by capturing performance data, usage, and monitoring dependencies before and after migration.

Migrate and modernize

This phase is the actual implementation of migration and validation of your applications. There are specific sets of services that AWS offers to streamline and simplify the process of three main approaches, such as the following:

- Migrating applications to AWS

- Modernizing applications that are already running on AWS

- Rehosting or replatforming your applications to AWS

AWS Application Migration Service provides capabilities to quickly lift and shift a large number of servers from physical, virtual, or cloud infrastructure to AWS. MGN can automate the process of conversion and allows you to realize the benefits of migrating applications to the cloud without changes and with minimal downtime. Critical applications such as SAP, Oracle, and SQL Server can be easily migrated by launching non-disruptive tests and continuously replicating source servers, which means little to no performance impact.

MGN is an agentless service that makes it easier and faster to migrate thousands of on-premises workloads to AWS from a snapshot of an existing server. Here are some of the features of MGN:

- **Improved version of CloudEndure Migration**: Offering direct integration with the AWS management console.

- **Supports migration**: Supports the migration of a variety of physical servers, VMware, vSphere, and Microsoft Hyper-V to AWS EC2 instances between AWS Regions or between AWS accounts.

- **A variety of OSs**: OSs across Windows and Linux are supported.

VMware Cloud on AWS delivers a faster, easier, and cost-effective path to the hybrid cloud while allowing your enterprises to modernize applications and accelerate the time-to-market and increased innovation. The collaborative partnership brought to light the new Amazon EC2 i3en.metal instances specifically for VMware Cloud on AWS, powered by Intel processors to deliver high networking throughput and low latency. This helps you to migrate data centers to the cloud for uses cases such as the following:

- Rapid data center evacuation

- Disaster recovery

- Application modernization

AWS Marketplace is a curated catalog of software solutions to support each phase of migration, such as data discovery, application profiling, workload mobility, security, and application modernization. Offerings such as single-click deployment and the flexibility to test software can help you accelerate the momentum and achieve business results quicker.

AWS Managed Services (**AMS**) helps you augment and optimize your operational abilities in both new and existing AWS environments. Services such as a library of automations, configurations, and runbooks are critical components to succeed in the cloud and enhance security and compliance by maintaining a growing repository of guardrails that you can leverage.

AWS Service Catalog is an approved set of IT services that you can use, including everything from **virtual machine** (**VM**) images, servers, software, and databases to complete multi-tier application architectures. You can centrally manage deployed IT services, including applications, resources, and metadata, to achieve consistent governance. As leaders of enterprises, you get to ensure compliance is met with corporate standards by meeting the granular control and configuration requirements. The life cycle of your IT applications can be centrally managed across the regions, and business stakeholders get up-to-date information on application content and metadata.

Amazon EC2 Spot

Spot Instances are the most popular option to drastically reduce expenditure on EC2 on-demand instances by as much as up to 90%. Enterprises of any size can leverage Spot Instances and they are best suited to stateless and fault-tolerant applications such as data analysis, batch jobs, and background processing that can survive possible interruptions.

> **Spot Instance**
>
> A Spot Instance uses unused EC2 capacity generally available for steep discounts compared to the on-demand price.

AWS offers this cost-effective choice to offset the loss of idle infrastructure and drive usage, which is the main reason for the cloud provider to offer massive discounts with the Spot Instances.

How does Spot work?

With Spot, you can use the same EC2 instance to reduce compute costs and improve application throughput. A set of unused EC2 instances called a *Spot capacity pool* for the same instance type and **Availability Zone** (**AZ**) will be offered at a *Spot price*, which is the current price of a Spot Instance.

Spot Instances can be instantiated in a couple of ways:

- Create a Spot Instance request by selecting a launch template or a custom AMI and configure the security and network access.
- Have AWS EC2 create one on your behalf.

When a request comes in via the *Spot Instance request*, the maximum price per hour is captured from the request based on what you requested. When the request exceeds the Spot price, Amazon EC2 fulfills your request given there is capacity.

The Spot Instance is terminated, stopped, or hibernated when EC2 needs the capacity back or the Spot price exceeds the maximum price for your request. EC2 automatically submits a persistent request after the instance associated with the request is terminated. A *Spot Instance interruption notice* is provided, which gives you a two-minute warning before the interruption.

Additionally, an *Instance rebalance recommendation* signal is sent to you if a given Spot Instance is at an elevated risk of interruption. Based on the signal, you can proactively rebalance your workloads across existing or new Spot Instances before having to wait for the two-minute interruption notice.

The Spot price is generally determined by the long-term trends in supply and demand for the EC2 spare compute capacity. You get to pay the Spot price that is in effect at the beginning of each instance hour for your EC2 instance that is running, and it is billed to the nearest second.

The advantages of Spot Instances include low prices, massive scale, ease of use, ease of automation, and integration with other AWS services such as AWS Batch, Amazon **Elastic MapReduce** (**EMR**), and so on.

Things to consider

Here are the things to consider when using AWS Spot Instances:

- Make sure your teams have a durable Spot allocation strategy and know when to use the Spot and when not. Having allocation strategies such as auto-scaling the Spot Fleets automatically will help ensure optimal performance when your workload traffic increases.

- Rightsizing your Spot Instances is one way to guarantee optimal performance and optimize the resources effectively. Using AWS CloudWatch to rightsize your instances and automation tools to create a repeatable process is recommended.

- Spot Instances are not always available and, as a result, you need a backup plan if you want to rely on the Spot Instance for long-term purposes. Running a combination of workloads, such as on-demand **Reserved Instances** (**RI**), can ensure that your operations continue running when there is an interruption.

Use cases

Spot Instances can be used to run many workload types that are distributed, scalable, and fault tolerant:

- **Containerized workloads**: Using Spot Instances to run containers helps reduce costs when compared to the price that you pay for on-demand instances.

- **Big data and analytics**: Using Spot Instances can fast-track your big data and ML workloads. Many companies find Spot Instances to be an optimal combination of cost savings, acceleration, and scale. Spot Instances can be used with Amazon EMR, Hadoop, or Spark for processing massive amounts of data.

- **High-performance computing**: Big compute workloads such as algorithmic trading and genomic sequencing can be accelerated using Spot Instances with AWS Batch.

- **Web services**: You can save up to 90% on web services and applications. You can scale tens of thousands of instances when you deploy EC2 Spot Fleet behind a load balancer.

- **Continuous Integration/Continuous Delivery (CI/CD) and testing**: You can automatically scale a fleet of Spot Instances when using Jenkins with the EC2 Spot.

- **Image and media rendering**: Manage workloads cost-effectively with near-limitless capacity and take advantage of **Bring Your Own License** (**BYOL**) for popular rendering and content creation software.

Best practices

Applying the best practices for your applications that are leveraging Spot Instances is the key to making sure that you are reaping all the benefits of Spot:

- Be flexible in choosing your instance types and the AZs where you can deploy your workloads. This will improve your chances of finding and allocating the required compute capacity. Having multiple options while requesting will put you in a better place to procure the compute capacity.

- Use capacity-optimized strategies to provision instances from the most available Spot Instance pools automatically. This improves the possibility of continuously running your Spot Instances.

- Use proactive capacity rebalancing to ensure workload availability and use your autoscaling group with a new Spot Instance before receiving the two-minute Spot Instance interruption notice.

- Use integrated AWS services to manage your Spot Instances and reduce overall costs. Spot can integrate with Amazon EMR, Amazon ECS, AWS Batch, Amazon **Elastic Kubernetes Service** (**EKS**), SageMaker, AWS Elastic Beanstalk, and Amazon GameLift.

Let us take a look at additional purchasing options for EC2 in the next few subsections.

Additional purchasing options

Amazon EC2 offers multiple purchasing options when it comes to optimizing the costs and running the workloads using the right purchasing option that suits your requirements. Before getting into the actual purchasing options, it is important to understand that the purchasing option you choose affects the life cycle of an instance.

> **Instance Life Cycle**
> The life cycle of an instance starts when it is instantiated or launched and ends when it is terminated.

Let us take a look at the available purchasing options.

On-Demand Instances

With On-Demand Instances, end users pay for compute capacity by the second. You get to have full control over the instance's life cycle, and you get to decide when to launch, stop, hibernate, start, reboot, or terminate it. There are no long-term commitments, and you pay only for the seconds that your On-Demand Instances are in the *running* state with a 60-second minimum.

Advantages

There are multiple benefits of using on-demand pricing, including the following:

- You get full control of the EC2 instances and their life cycle.

- It comes with a **Service Level Agreement** (**SLA**) at the Region level across multiple AZs and at the instance level that governs the EC2 instances. This ensures the uptime guarantee and improves the consumer experience.

- Pay-as-you-go pricing ensures no upfront costs to procure the On-Demand Instances.

- You are not required to make long-term commitments.

- Scaling the EC2 instances up or down is a flexible option with the On-Demand option.

- You get the option to experiment with different deployments and establish the most appropriate balance of EC2 resources that are needed to support your workloads.

- Versatile EC2 On-Demand types can support practically any workload.

Limitations

On-Demand Instances comes with a set of drawbacks that you need to consider:

- On-Demand Instances is comparatively more expensive than other options available to purchase for EC2 instances.

- The flexibility to scale also comes with the challenge of accumulated volumes of underutilized resources.

- The complex set of pricing options available within On-Demand Instances makes it difficult for beginners to navigate.

Use cases

You can opt for On-Demand Instances in the following scenarios:

- **Temporary usage**: Applications that don't require instances to run on a long-term basis can be set up in a short-term test environment mode and can use On-Demand Instances.

- **Unpredictable traffic**: If you are running applications on a trial mode and need a pay-as-you-go model to allow you to test out the computing, then On-Demand can be a good choice.

- **Urgent deployments**: Workloads that need an urgent deployment can leverage On-Demand Instances almost instantly.

- **Applications running on a given schedule**: If you know you are only going to use a particular instance part-time – say, eight hours a day, six days a week – then using On-Demand Instances can help you to experiment and pick the most appropriate for your performance needs.

- **Business operations**: With steady growth, you can leverage on-demand instances due to the flexible scaling that it offers.

Next, let us take a look at other pricing models that offer varied pricing options.

Savings Plans

Savings Plans is a pricing option that offers lower prices compared to On-Demand Instances but requires specific usage commitment for a one- or three-year period. The pay commitment options are **All upfront**, **Partial upfront**, or **No upfront**, so you get to choose the optimal option:

All upfront: You get the largest discount with this payment plan.

Partial upfront: You get a 1%–10% lower discount than the **All upfront** option.

No upfront: You get a smaller discount compared to the previous two options but you can free up your capital to invest in other efforts.

The usage commitment is measured in dollars per hour regardless of the instance family, size, OS, tenancy, or AWS Region. Savings Plans comes in two types: **Compute Savings Plans** and **EC2 Instance Savings Plans**:

- Compute Savings Plans

  Prices for this option type are up to 66% off On-Demand Instances, and the plan automatically applies to EC2 instance usage irrespective of the other attributes that I stated previously. This option drastically simplifies things as you do not have to go through time-consuming efforts to figure out the configurations before you make a commitment

- EC2 Instance Savings Plans

  EC2 Instance Savings Plans offers up to 72% savings for individual instance families in a given Region. This helps in reducing your costs on a selected instance family in that Region regardless of the AZ, size, OS, or tenancy. Savings Plans provides the flexibility to change your usage between instances within a family in that Region

  For example, you can change your instance type from c5.xlarge on Linux to c5.3xlarge running on Windows and automatically benefit from the Savings Plans prices.

Here is a quick comparison between the two Savings Plans options:

	Compute Savings Plans	**EC2 Instance Savings Plans**
Savings off On-Demand Instances	66%	72%
Change instance family	Yes	No
Change OS	Yes	Yes
Change tenancy	No	Yes
Change Region	Yes	No
Change instance size	Yes	Yes
Change compute options	Yes	No
Supported AWS compute options	EC2, Fargate, Lambda	EC2

Let us take a look at more details on Savings Plans so you can do a deeper evaluation and determine what works best for your application environment.

Advantages

Savings Plans is simpler and easier to manage and come with its own sets of benefits:

- Savings Plans reduces the cost of compute usage across AWS Regions without having to exchange or modify instances.

- Savings Plans requires no monitoring, and the discounts are applied to the instances that provide you the greatest discounts first.

- The ability to penetrate sparsely used AWS Regions makes Savings Plans a great choice where usage is unpredictable.

Limitations

Here are some challenges that you need to know about:

- Three-year commitments can be tricky because AWS Savings Plans cannot be sold in the AWS Marketplace. This prevents you from upgrading to newer generation types that AWS launches for better performance and savings.

- Savings Plans is only applicable to a few AWS offerings.

The discounts are not as high for Windows as they are for Linux, which makes the blended use of deployments less compelling.

Use cases

Every enterprise would like to save money and stop struggling to manage costs. Let us take a look at some of the most common scenarios where you can benefit from Savings Plans:

- Steady-state workloads running on AWS can benefit from the forecastable nature of Savings Plans where long-term commitment to AWS does not impact your applications.

- If you are a start-up and know the bare minimum to run your platform on AWS, you can start by committing to a small amount of X usage for an affordable price.

If you are launching and moving workloads onto AWS and are highly unlikely to switch to other cloud providers, Savings Plans will provide a baseline that can help you save money.

Reserved Instances (RI)

As the name suggests, you can invest in reserved capacity and save up to 75% over an equivalent on-demand compute capacity. Reserving the key resources and capacity for a period of one to three years for a specific AZ inside a Region to gain significantly lower rates compared to the on-demand rates is the idea behind having RI.

When you purchase RI instances, you can match the selected attributes of an existing or new On-Demand Instance and opt to get charged at the discounted price. The discounts are automatic and don't affect the running state of any existing instances.

Let us consider a scenario; the following diagram shows the details of the scenario on how to use RI:

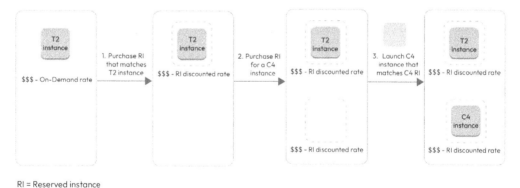

Figure 7.3 – RI purchasing overview

For the scenario, you have an existing On-Demand Instance (**T2**) running on your AWS account, paying the price of On-Demand. Once you purchase an RI instance that has the same attributes as this T2 instance, the discounts are applied immediately. In addition, you also purchase an RI instance for a C4 instance type, but you do not have any running instances of the same type. When you launch the C4 instance, the billing benefits are automatically applied.

Some of the key attributes that determine the RI pricing are the instance type, Region, tenancy, platform, term commitment (one-year/three-year), payment options (All upfront, Partial upfront, No upfront), and offering class (Standard/Convertible).

Offering classes for RI have the following characteristics:

Standard RI	Convertible RI
One-year to three-year commitment	One-year to three-year commitment
You cannot change the instance family	You can change the instance family
You cannot change the OS	You can change the OS
You cannot change the tenancy	You can change the tenancy
You can change the AZ	You can change the AZ
You cannot change Regions	You cannot change Regions
Can be sold in the RI Marketplace	Cannot be sold in the RI Marketplace

Advantages

Some of the advantages of using RI are as follows:

- The main advantage is the cost savings, where RI let you reduce costs that you are using already.

- You get to reserve the capacity in your preferred AZs across multiple geographic locations enabling you to take full advantage of auto-scaling whenever you experience any traffic spikes for your applications.

- When you reserve the compute capacity for a specific AZ, your applications can benefit from low latency and run independently of other AZs.

- The concept of RI is especially beneficial in times of crisis of exhausted capacity in a given AZ. If there is an infrastructure issue due to a natural calamity or a power failure, you will get a front spot in the queue during the shift to a different infrastructure. It is similar to insurance coverage in case you exhaust your capacity.

- The flexibility to move to a different AZ enables you to modify the network platform and instance size of the reservation to get another size of that instance type for no extra charge.

- The pricing benefits are shared when the purchasing account is billed under a consolidated billing payer account. This is beneficial for companies that have multiple functional teams or groups and want to benefit from the normal RI logic.

Limitations

RI comes at a price where you may have to consider the following risks before moving forward:

- When committing to specific configurations or levels of usage, any requirement changes cannot happen while the contract of an RI instance is in place. It is important to make accurate estimates for your workloads, and that is often daunting.

- Signing a contract for RI means you will be running your workloads on AWS for that period of time. Switching to different vendors comes with high costs.

- If your team requires fewer resources than expected, you are committed to a given capacity and have paid for it. Therefore, there is a chance that you'll end up with unused capacity.

- Less flexibility in scaling or options to configure resources in multi-Regions makes it difficult to plan.

Use cases

RI is most suitable for the following scenarios:

- If you have production workloads or applications that require the compute capacity to be available 24x7.

- Mission-critical applications running on multiple AZs to ensure high availability can leverage the reserved compute capacity in the form of RI.

- Workloads with a steady state and predictable performance can benefit the most from RI.

Dedicated Hosts and Dedicated Instances

Dedicated Hosts and Dedicated Instances are options for end users to leverage EC2 instance capacity in a fully dedicated manner. Both options provide you with ways to launch instances to Dedicated Hosts with resources that cannot be used by any other customers.

While there are no technical, performance, security, or physical differences between these two options, the real difference is in the visibility of the physical attributes:

	Dedicated Host	**Dedicated Instance**
Billing	Per-host billing	Per-instance billing
Visibility	Provides visibility into the number of sockets and physical cores	No visibility
Host and instance affinity	Dedicated physical server over a specific period of time	Not supported
Target instance placement	Visibility and control over the placement of instances on the physical server	Not supported

	Dedicated Host	**Dedicated Instance**
Automatic instance recovery	Supported	Supported
Bring your own license (BYOL)	Supported	Not supported
Per-instance billing	Not supported	Supported
Per-host billing	Not supported	Supported
Increase capacity through allocation request	Not supported	Supported

While Dedicated Instances are very valuable from a compliance perspective, Dedicated Hosts give you visibility into the physical host to let you use your own Windows Server, SQL Server, or **Red Hat Enterprise Linux** (**RHEL**) licenses that are provided on a CPU core basis.

On-Demand Capacity Reservations

On-Demand compute capacity can be reserved in a specific AZ for any duration using this option. You get the ability to create and manage the reservations independently from the billing discounts coming from Savings Plans or RI.

The main idea is to ensure that you always have access to the EC2 capacity when you need it and for as long as you need it. You get to select the AZ, number of instances, and instance attributes when you create a capacity reservation.

The following table highlights the key differences between Capacity Reservations, RI, and Savings Plans:

	Capacity Reservations	**Zonal RI**	**Regional RI**	**Savings Plans**
Term	No minimum commitment required	One to three years	One to three years	One to three years
Capacity	Per AZ	Per AZ	No capacity required	No capacity required
Discount	No discounts	Supported	Supported	Supported
Instance limits	Limits per Region apply	20 per AZ	20 per Region	No limit

While the purchasing options are exhaustive, knowing the details of building a cost-effective environment on the cloud helps you address the **return on investment** (**ROI**) aspect of migrating and modernizing on the cloud.

In the next section, we will discuss how Auto Scaling works on EC2, which is a core concept for any AWS-specific offering.

Amazon EC2 Autoscaling

EC2 Auto Scaling is a fully managed service where you can maintain application availability and dynamically add or remove EC2 compute according to the configurations you define. Dynamic scaling enables your workloads to respond to changing needs and predictive scaling helps assign the right number of EC2 instances that your application needs at any given point in time.

For example, Netflix leveraged the broadest and deepest set of EC2 platforms to meet the growing viewing demands and increased member growth during the unprecedented times of the COVID pandemic. Netflix was able to scale the computing capacity needs and rapidly scale up to meet the demand spikes.

EC2 Auto Scaling helps you maintain instance availability whether you are running one or a thousand instances. Automatically detecting impaired EC2 instances and replacing the instances without interrupting your workload operations ensures high availability.

Here are some of the features of EC2 Auto Scaling that make it very favorable for modern applications to leverage and take benefit from:

- Scale in automatically when demand increases and scale out by terminating the unnecessary EC2 instances automatically and save costs when demand subsides.

- Control when and how to scale dynamically based on the Amazon CloudWatch metrics. You can also define a schedule based on the predictability of your workload traffic and set up CloudWatch alarms to trigger EC2 Auto Scaling actions.

- Fleet management automation can help maintain the availability of your applications by automatically replacing unhealthy or unreachable instances.

- Predictive scaling can provide predictions on future traffic, including patterns on traffic spikes, and helps you provision the right number of EC2 instances.

- Supports multiple instance attributes and lets you provision and scale instances automatically across purchase options, AZs, and instance families to optimize scale, performance, and cost.

EC2 autoscaling integrates with AWS **Identity and Access Management** (**IAM**) to ensure granular access controls and resource-level permissions. With all the characteristics we discussed previously, fault tolerance improvements and increasing your application availability while lowering costs are guaranteed.

Amazon Lightsail

Amazon Lightsail offers pre-configured and easy-to-use virtual private servers, containers, storage, and databases at a low cost.

Here are some of the use cases of Amazon Lightsail:

- Launch simple web applications using pre-configured blueprints such as WordPress and PrestaShop using LAMP.
- Create custom websites and host static content, and connect your content to an audience around the globe with just a few clicks.
- Launch and build small business applications such as file storage and sharing and financial and accounting software.
- Spin up test environments and development sandboxes easily where you can innovate and experiment with new ideas, risk-free.

There are many use cases where customers can leverage Lightstail and accelerate their digital transformation. For example, a gaming company was easily able to innovate quickly and cost-effectively with just a small IT staff. The company used Lightsail to host their platform, accelerate service to its customers, and connect with gaming servers around the world. With Lightsail's global network, the company was able to speed up and enhance its end user experience while not worrying about costs.

We discussed in the previous chapter the benefits of using microservices and how microservices have become the industry standard for many enterprises. In the next section, we will look at the concept of containers and how to run containers on AWS.

Diving deep into containers

In the past few years, container technology has rapidly grown in popularity, and technology leaders have embraced this technology in their business's application modernization strategy.

The rise of containers

Having a diverse portfolio of applications, cloud, and on-premises-based infrastructure makes it difficult for any business to transform. Container technology redefines how your applications need to be packaged such that they can run with isolated dependencies. Containers have changed the dynamic of application modernization and how modern IT infrastructure can be effectively managed in the cloud. The main idea behind this technology is for businesses to be able to run portable software without having to worry about the computing runtime environment: application and its dependencies, libraries or any other configuration files needed to run them.

> Container
>
> A container is a lightweight and stand-alone executable package of application code, runtime, software libraries, and configuration files required to operate an application.

Although container technology has been part of the Linux world since the 1980s, the introduction of **Docker** (open source **platform-as-a-service (PaaS)**) brought remarkable improvements from 2013 onward in container technology. The trend toward containerization has grown dramatically since then, and today, as per the *Container Infrastructure Market Assessment*, 76% of enterprises are running their mission-critical workloads as containers in production. The remaining have plans to use them in the near future.

Docker came up with an approach to package the tools required to build and launch containers in a more streamlined and simplified manner. Docker has provided phenomenal capabilities right from its inception that have changed how applications can be created or managed at scale. Capabilities such as an easy-to-use **graphical user interface (GUI)** and support for multiple applications using different OSs led to 100 million downloads of Docker within a year of its inception. The simplification of deployments and portability of the reusable images made it extremely flexible for developers to deploy, replicate, move, and back up any workloads in a streamlined way.

Docker established the following *three* core concepts that can transform the way you ship your software applications:

- Using immutable container images that can be easily shipped across several open standard platforms.

- Creating central registries to store these container images and leveraging these creates strong governance.

- Having a container runtime engine with clearly defined responsibilities can streamline the building process of containers.

Docker revolutionized use cases such as big data analytics, application frameworks, application infrastructure, application services, databases, messaging services, monitoring, deep learning, storage, DevOps tools, and many more. Irrespective of the programming language you use to develop your applications and platforms, the standard mechanism of packaging the application and running it seamlessly across multiple environments empowers developers.

Comparing traditional deployments versus virtualization versus containers

When it comes to the way we deploy and ship software applications, we have come a long way. Let us take a quick look at the evolution, starting with traditional deployment.

Traditional deployments

The traditional style of deployment and change management processes is often measured by service stability. Organizations run applications on physical servers and use application package files from an artifact source such as a **Web Application Resource (WAR)** file and deploy them on a runtime environment using the target host. Let's take a look at the characteristics in detail and its challenges.

Characteristics of traditional deployments

- Applications are hard-wired to the environment of the physical servers.

- The dependent tools and libraries needed to be installed on an existing OS.

- The environment is mutable because you are allowed **Secure Shell** (**SSH**) protocol access.

- Deployments involve a series of sequential manual steps that have to be executed in the right order of events.

Challenges with traditional deployments

- Manual deployment processes can lead to multiple potential untracked breaking changes.

- The inability to define resource boundaries for applications in the physical layer.

- Cannot scale applications, and long downtimes can lead to business risks.

- Expensive to maintain and track the manual deployments.

- Overutilization of resources and cascading failures of application crashes can bring down the entire physical server.

- Managing service-oriented architecture complexity for platforms that leverage microservice-based architectures can result in complexities of cross-service dependencies for deployment.

- Configuration at scale, when you have dependent services across multiple environments, gets challenging. You have to track the number of configuration points that you need, and relying on manual efforts to configure each of these can be cumbersome. Another pain point is managing customer-specific configurations.

- Traditional deployments are typically linear pipelines that are tightly coupled and can be problematic to standardize or reuse. This can be problematic for patching, upgrading, and maintaining the asset dependencies and database schemas.

Virtualized deployments

Virtualization was introduced to solve the challenges of traditional deployments. Let's take a look at the characteristics in detail and its challenges.

Characteristics of virtualized deployments

- Virtualization allows the applications to run on multiple **VMs** using a single physical server's CPU.

- Each VM will have an OS, required binary files, and libraries to support the operation of the application.

- Supports effortless updates to applications thereby attempting to solve the scaling issues.

- Allows better utilization of resources.

- Isolates applications between VMs.

Challenges of virtualized deployments

- Each workload needs a complete OS to run, which increases the image size, and OS images are generally heavyweight. A bigger OS footprint means increased times to spin up apps and slow boot-up processes.

- Security updates increase operational overheads.

- As the number of VMs proliferates, IP addresses can be lost or the images can be assigned to the wrong **virtual local area network (VLAN)** causing a lapse in application availability.

- Additional costs from implementing OS virtualization, procuring new hardware, and software licenses to improve availability, performance, and management.

- Managing VMs can be a struggle, especially when it comes to managing the host system and the hypervisor.

The following diagram depicts the differences between each type of deployment:

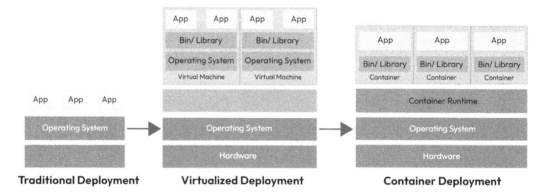

Figure 7.4 – Traditional versus virtualized versus container deployments

Containerized deployments

The process of pushing containers to the target environment, cloud, or on-premises server is called container deployment. Containers are the core building block of modern software development and are particularly applicable to digital transformation goals.

Let us take a look at the characteristics and benefits of containerized deployments.

Characteristics of containerized deployments

- Containers break down the wall between developers and operations. The common packaging tools allow your application to become portable across any environment.

- They abstract the OS layer and package code and dependencies together into lightweight immutable units.

- They take less memory and disk space and, as a result, boot quickly.

- There is minimal wastage of resources given the containers use the underlying hardware.

- Containers are process-isolated and do not require a hypervisor.

Benefits of container deployments

- **Speed**: Due to the inherent nature of containers being lightweight, the operational effort required to ship the code to production, including areas such as infrastructure provisioning and testing, is simplified and results in faster and more frequent deployments.

- **Agility and flexibility**: Containers support business goals that are fluid and evolving. Changes are quickly accommodated with the support of microservices design and can also lead to improved security control.

- **Resource optimization**: Containers being lightweight demands less load on system resources, and multiple applications can share the same OS.

- **Run anywhere**: Containers are abstracted away from their underlying OS and infrastructure, meaning the code will behave in the same manner no matter where it is running.

There are a variety of tools that containers can be deployed with. As we discussed previously, Docker is one popular container platform. Following the popularity of Docker, *Google* introduced **Kubernetes**, an open source project whose adoption and popularity exploded in the cloud. Kubernetes is a portable, extensible, open source platform for managing containerized workloads and services. Kubernetes is now maintained by the **Cloud Native Computing Foundation** (**CNCF**), a vendor-neutral hub for many fast-growing open source projects. Features such as service discovery, load balancing, self-healing, and horizontal scaling offered by Kubernetes enable automated deployments and simplify containerized workload management.

OpenShift by Red Hat, Rancher, and Mesos are some of the popular container orchestration open source platforms that can be leveraged by organizations to directly start containerizing their workloads.

How to run containers on AWS

Having a container management strategy is important when it comes to containerizing your workloads at scale on AWS. There are 17 ways to run containers on AWS as of 2022 when it comes to utilizing various AWS services such as AWS Proton, AppSPACERunner, Lightsail containers, EKS, ECS, **Red**

Hat OpenShift Service on AWS (ROSA), CodeBuild, Fargate, App2Container (A2C), Copilot, Elastic Beanstalk, and AWS Lambda.

These container offerings fall into the following categories to enable a successful container management strategy, which we will review in more detail in the next few subsections:

- **Orchestration**: The automation of the deployment, management, scaling, and networking of containers.

- **Registry**: The catalog of repositories used to store and access container images.

- **Tools**: The **command-line interface** (**CLI**) tools enable the quick launch of containerized applications.

Container registry

Container registries play a vital role in a successful container management strategy.

A **container image** holds your application code and all its dependencies that are required for it to run.

A **container repository** is used to store container images for setup and deployment.

A **container registry** is a collection of repositories to store container images.

Image scanning allows you to identify vulnerabilities within your container images.

The following diagram shows how container images go through a three-step process – build, store, and run:

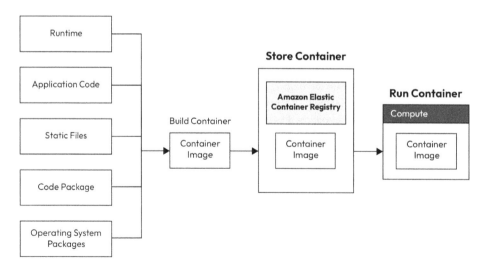

Figure 7.5 – The build, store, and run process

Once a container image is stored in a registry, other hosts can download it from the registry server.

Public container registries are generally the fastest and easiest way to access open sourced images and are typically ideal for organizations that have more freedom to leverage such images. Public registries are less secure than private registries.

Private container registries are set up for organizations to have complete control over the registry and be more secure when it comes to implementing security measures. If an organization's priority is security, then a private registry should be implemented so that you get the ability to scan the images for any vulnerabilities, digitally sign images to ensure the images are trusted, and use authentication methods for assigning **role-based access control (RBAC)**.

There are many registries available in the market, and choosing one can become a difficult task. AWS offers **Elastic Container Registry (ECR)** with the following features:

- Docker support allows you to use the Docker CLI commands to communicate with ECR and push, pull, list, and tag the container images.

- The public artifact gallery lets you discover and use container software and open source projects, and developers can access it from the public gallery.

- AWS Marketplace container software that you buy can be stored in ECR.

- Access control and encryption allow you to use IAM to control who has access to the container images and specify roles for different users. The container images can be transferred to and from ECR via HTTPS and also support encryption at rest using Amazon **Simple Storage Service (S3)** server-side encryption.

- Automatic integration with Amazon container orchestrators such as Amazon ECS allows you to retrieve and store the required images for your applications.

- Image scanning is a crucial part of container security for developers, security operations engineers, and infrastructure admins. ECR supports two types of scanning: static scanning and dynamic scanning. Static scanning enables developers to detect vulnerabilities before the container is launched. Dynamic scanning is executed in a runtime environment to detect vulnerabilities for containers running in test, **quality assurance (QA)**, or production environments.

The following diagram depicts the end-to-end process of storing and pulling the container images for deployment using Amazon ECR:

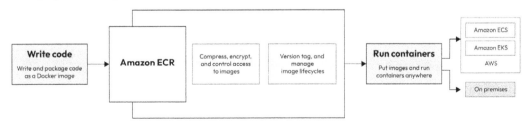

Figure 7.6 – Amazon ECR

Use cases

You will find Amazon ECR fits your needs in the following scenarios:

- If you want to improve the security of your container images and ensure integration with the Amazon container orchestrator, your enterprise can benefit from the broad range of OS vulnerabilities.

- For enterprises that need a fully managed Docker container registry and have a reliable way to replicate images.

- Additional performance benefits to keep all the network traffic within the AWS network and resolve rate-limiting issues from pulling images from a public registry can ensure that your enterprise can leverage ECR as a suitable tool.

In the next section, let us take a look at the container orchestrators that Amazon offers.

Orchestration

Enterprises find it very difficult to deploy and manage containers at scale and often end up building container management tools from the ground up. With microservices and containerization, deploying a self-contained deployable unit across different environments is supported but managing its life cycle can often be a time-consuming task.

Container orchestration is the process of automating the tasks pertinent to the life cycle of containers, such as the provisioning, configuration, scheduling, resource allocation, deployment, scaling, load balancing, monitoring, management, and networking of containers. The following diagram shows the typical steps included in container orchestration:

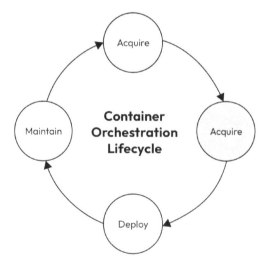

Figure 7.7 – Container orchestration

The manual pains of deployments are mitigated and efficiency is maximized with container orchestration along with many other benefits, as shown in the following diagram:

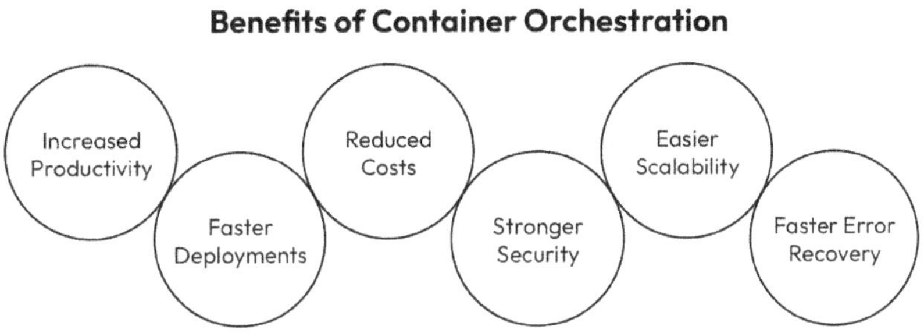

Benefits of Container Orchestration

Figure 7.8 – The benefits of container orchestration

Container orchestration tools are based on declarative infrastructure programming, where you define the state that you want the orchestrator to accomplish. The output is achieved by the orchestrator without you having to worry about the process that is involved in making it happen. Each container orchestration platform is implemented differently, so let us get started with understanding what AWS has to offer in terms of container orchestrators.

Amazon Elastic Container Service (ECS)

ECS is a fully managed container orchestration service that helps you to deploy, manage, and scale containerized applications easily. Its deep integration with the AWS platform, such as autoscaling, load balancing, IAM, networking, logging, and monitoring, makes it a favorable option for many who are already operating AWS.

ECS leverages the following core constructs to run applications:

- AWS ECR to manage, store, and control access to container images.

- An ECS cluster is a logical grouping of tasks and containers using different types of infrastructure, such as EC2 instances or Amazon-managed instances.

- Task definitions define the ECS application required to run the containers, typically stored in JSON files.

- ECS service scheduler to schedule task instantiations and run inside the ECS cluster infrastructure. The service schedulers typically execute tasks on a cron-like schedule or a custom schedule. It handles various task placement strategies for selecting instances and minimizing unused CPU or memory.

How does ECS work?

To understand how Amazon ECS works and how you can use it, it helps to understand the basic blocks of ECS and how they fit together:

- **Container instance:** In the context of Amazon ECS, a container instance is an EC2 instance running an agent that has been registered to your cluster. Container instances use any AMI that has a modern Linux distribution with an agent and Docker daemon with runtime dependencies running on it. AWS simplified the process by creating an Amazon ECS-optimized AMI just for this purpose.

    The following diagram depicts all the essential blocks of Amazon ECS:

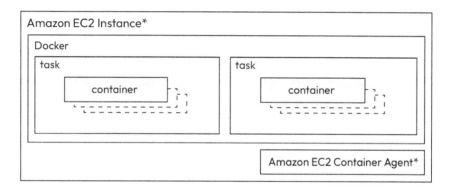

Figure 7.9 – Amazon ECS building blocks

- **Cluster:** An ECS cluster is a grouping of container instances in a specific Region and can span across multiple AZs.

- **Agent:** The ECS container agent is responsible for handling the communication between the scheduler and your instances. Tasks such as registering your instance into a cluster are taken care of by the agent.

- **Task:** Tasks are a logical grouping of 1 to N containers that run together on the same instance. It is a basic frontend unit that comprises a web server, an application server, and an in-memory cache.

- **Task definition:** Think of a task definition like an architectural plan for a city. It defines how the containers interact and container CPU and memory constraints. The task definition typically has the parameters shown in the following diagram:

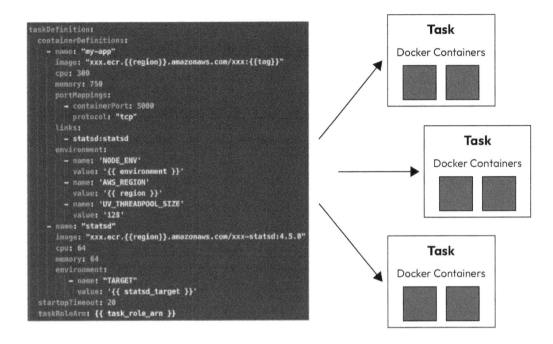

```
taskDefinition:
  containerDefinitions:
    - name: "my-app"
      image: "xxx.ecr.{{region}}.amazonaws.com/xxx:{{tag}}"
      cpu: 300
      memory: 750
      portMappings:
        - containerPort: 5000
          protocol: "tcp"
      links:
        - statsd:statsd
      environment:
        - name: 'NODE_ENV'
          value: '{{ environment }}'
        - name: 'AWS_REGION'
          value: '{{ region }}'
        - name: 'UV_THREADPOOL_SIZE'
          value: '128'
    - name: "statsd"
      image: "xxx.ecr.{{region}}.amazonaws.com/xxx-statsd:4.5.0"
      cpu: 64
      memory: 64
      environment:
        - name: "TARGET"
          value: '{{ statsd_target }}'
  startupTimeout: 20
  taskRoleArn: {{ task_role_arn }}
```

Figure 7.10 – The anatomy of a task definition

The Docker image to use, the CPU and memory required for each task or container within the task, the launch type to use, the Docker networking mode, and the IAM role that your tasks use are some of the important parameters.

- **Scheduler:** The scheduler is part of the hosted orchestration player that decides which containers to run where according to a number of constraints.

- **Service:** A service ensures that your tasks are running. The service is the construct that defines, at any given time, that N tasks are running using the task definition.

The following diagram depicts a detailed illustration of how ECS works:

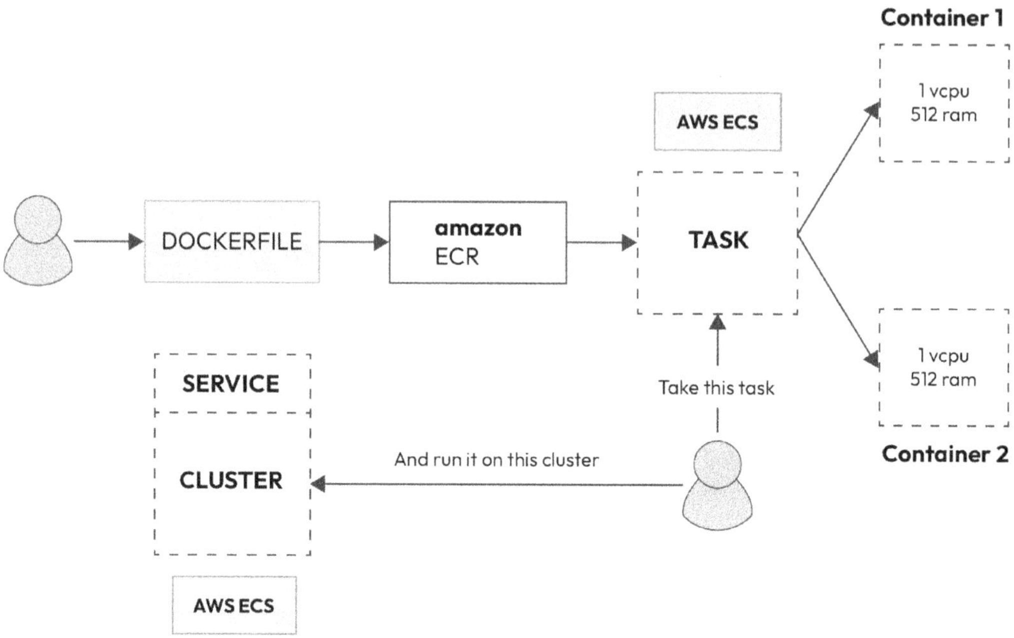

Figure 7.11 – An overview of how Amazon ECS works

Let us take a look at how ECS works when you set up an ECS cluster:

- **Define your Dockerfile**: In order to run your tasks, you need to define a Dockerfile, which typically contains references to the source code, dependencies, port mapping configurations, and startup scripts.

- **Build and push the image to ECR**: Once the Dockerfile is ready, you can build the file into an image. A container image is a lightweight executable package of software that is required to run an application. Upload the Dockerfile to Amazon ECR, where all the images are hosted and can be used to launch onto the ECS containers.

- **Define the ECS task**: Once the Docker image is uploaded to ECR and is given a name, the next step is to define the ECS task. The ECS task is a set of guidelines that tells ECS how to spin up or down the containers. A task can contain one or multiple containers to support various functionalities in an application.

- **Set up the ECS cluster**: A cluster is an abstract resource form to set up instances. The ECS agent communicates with the cluster and handles requests to launch new software or additional configurations for the cluster. The agent is also responsible for ensuring communication is seamless between the cluster and EC2 instances.

Features of ECS

Let us take a look at the capabilities of ECS to see where it fits in the best:

- **Simplified deployment**: With integrated Docker support, you can define and run multi-container applications easily. ECS also supports Windows containers, which makes container management easy to run on a Windows OS. The task definition (JSON template) lets you specify the configurations on CPU, memory, and how you want to run the containers.

- **Scheduling capabilities**: You can run multiple scheduling strategies to place containers across your clusters based on resource availability requirements. This makes it suitable for applications such as batch jobs, long-running applications, or daemon processes.

- **Native integration**: There is a wide range of AWS services that ECS integrates with and you get to leverage features such as load balancing, VPC, IAM, and storage volumes via Amazon **Elastic Block Storage** (**EBS**). ECS also integrates with AWS Cloud Map, a cloud resource discovery service that makes it easier for your containerized services to discover and connect with each other. Integration with service meshes such as AWS App Mesh makes it easy to configure end-to-end visibility across your containers. Monitoring is seamless through integration with Amazon CloudWatch.

With application modernization as the core of your digital transformation goals, ECS plays a crucial role in streamlining your app containerization efforts.

Use cases

It is important to choose container tools that focus on reducing operational overhead and scale by migrating existing container applications or building platform-independent cloud solutions by migrating to managed container services such as ECS to gain additional agility benefits and move away from technical debts. Let us take a look at some of the application modernization use cases for ECS:

- **Building secure microservices**

  A combination of Amazon ECR and ECS provides you with the ability to transfer container images over HTTPS for automatic encryption at rest. You can configure policies to manage granular permissions and control access to the images using AWS IAM users and roles. This granular control lets you limit access and honors the principle of "least privilege," which helps improve the overall security of your applications.

- **Faster time from ideation to market**

  If your enterprises are struggling with a lack of automation and you are manually deploying containers, ECS simplifies the task of running Docker workloads securely, at scale, and integrating them with other AWS services. There is no **control plane** to manage and you can leverage all your existing AWS skills and knowledge. Many enterprises are opting for smaller software delivery teams that are cross-functional and operate in a two-pizza team model.

> **Two-Pizza Rule**
>
> The two-pizza rule was instituted by Amazon's founder Jeff Bezos. The intention of this rule is very simple yet focuses on two goals: efficiency and scalability. Every internal team should be small enough that it can be fed with two pizzas. The small team acts as a business owner for a specific product line and works together to achieve the company's larger goals.

For such modern enterprises, ECS is a great fit given its simplicity in getting the applications up and running. The fully managed platform offered by ECS helps you focus on migrating your applications rather than worrying about any platform-specific issues.

- **Improved compatibility**

    Many enterprises experience unplanned work that typically transpires from environment variations during deployments, often leading to firefighting. Having a container-based pipeline can eliminate such issues and the lightweight nature of the containers makes it easy to replicate the deployments in any environment. ECS's compatibility with the AWS ecosystem brings an agile workflow and reduces unpredictability in terms of planning the work and spinning up applications. Additionally, automated deployment workflows can be set up using AWS CodePipeline and AWS CodeCommit to build the Docker image on AWS CodeBuild. Integration with **Application Load Balancer** (**ALB**) and automatic scaling makes it ideal for organizations of any size to streamline their DevOps teams and improve the process flows.

- **Scale ML models**

    With ECS, you can continuously deploy secure and scalable container-based deep learning and ML models along with prediction APIs. When deploying ML algorithms, it is important to consider factors such as memory, CPU, scale, and latency. The cluster management offered by ECS makes it easy to schedule the containers in a secure, scalable, and reliable environment, making it ideal for high and dynamic **transaction per second** (**TPS**) use cases. ECS provides the flexibility to optimize the prediction infrastructure cost-effectively.

Amazon Elastic Kubernetes Service (EKS)

Amazon EKS launched in 2017 and is a certified Kubernetes conformant orchestrator that lets you easily run your existing Kubernetes workloads on AWS. This certified Kubernetes offering enables you to migrate existing Kubernetes deployments and configurations without any changes. EKS was introduced to cater to customers that have Kubernetes as the core of their IT strategy and run millions of containers on AWS but face the challenges of operational overhead due to its self-managed nature. EKS aims to simplify processes such as building, securing, operating, and maintaining Kubernetes clusters.

With its deep integration, EKS leverages the AWS ecosystem of services and features to enable the container workloads to achieve performance, reliability, and availability at scale.

Amazon EKS manages open source Kubernetes clusters for you, allowing you to leverage all the existing plugins and tooling available from the Kubernetes community. EKS is built based on a shared responsibility model, as shown in the following diagram, where the control plane nodes are managed by AWS, and the **worker nodes** are your responsibility:

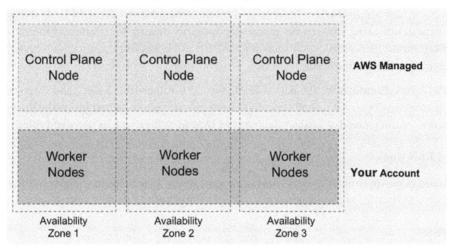

Figure 7.12 – EKS shared responsibility model

The Amazon EKS control plane consists of two API server instances and three *etcd* instances running across three AZs within an AWS Region for high availability. etcd is a key-value store to persist all the cluster data, such as configuration information, service discovery, scheduler coordination, and so on.

Features of EKS

Let us take a look at the Kubernetes-specific aspects applicable to EKS workloads running on AWS:

- **Managed control plane**: The control plane is managed by AWS and runs across multiple AZs of AWS in a given Region. The etcd persistence layer and the API servers are managed by the control plane. The self-healing nature of the control plane to replace the unhealthy control plane nodes makes it easy for you to focus on worker nodes and not worry about their availability.

- **Managed node groups**: The nodes for your EKS cluster can be easily created, updated, scaled, or terminated. The nodes can run EC2 instances and also leverage Spot to optimize costs.

- **Integration with AWS Cloud Map for service discovery**: Service discovery is important to ensure your application is available and discoverable. AWS Cloud Map lets you use external DNS and open source Kubernetes connectors to automatically propagate internal service locations to the service registry in Cloud Map.

- **Integration with AWS App Mesh for service connectivity**: Having a service mesh enables you to standardize communication across your microservices. Integration with AWS App Mesh gives you end-to-end visibility along with defining traffic routing and adding security features such as **Mutual Transport Layer Security (mTLS)** encryption and logging.

- **Open source compatibility**: There is an incredible community creating tools to improve the experience using Kubernetes given its popularity among the platforms for container orchestration. Open source tools such as CoreDNS and kubectl are compatible with Amazon EKS to supercharge your usage.

- **AWS IAM Authenticator**: The native RBAC system is Kubernetes RBAC, and Amazon EKS integrates RBAC with AWS IAM. AWS IAM Authenticator is a construct inside the Kubernetes cluster's control plane that enables the usage of IAM identities such as users and roles.

How does EKS work?

To understand how Amazon EKS works and how you can use it, it helps to understand the basic blocks of EKS and how they fit together:

- **Clusters**

    Clusters have two main components: the control plane and worker nodes.

    The EKS control plane runs on dedicated EC2 instances managed by Amazon to provide API endpoints accessible by your applications. The control plane is mainly responsible for managing the Kubernetes master nodes, such as the *API server* and *etcd*.

    The worker nodes run on EC2 instances managed by your organization. A unique certificate file gets created for each cluster. These nodes use the API endpoint to connect to the control plane via a certificate file.

- **Nodes**

    A node in EKS signifies an EC2 instance that Kubernetes Pods can be scheduled on. These Pods connect to the EKS clusters using the API endpoint. Nodes are grouped into node groups, which have an instance type, AMI, and IAM role. Each node in a node group can have a different type of instance with a different IAM role.

    There are three types of worker nodes when it comes to EKS: managed node groups, self-managed node groups, and Fargate. The managed node groups come with automated life cycle management where you can create, update, or shut down nodes automatically with one operation. The EKS cluster can schedule Pods in any combination of these three nodes.

- **Networking**

  EKS operates in a VPC where all the resources are deployed to an existing subnet in an Amazon Region. The control plane creates a VPC **Container Network Interface (CNI)**, a plugin that connects to the EC2 or Fargate instances. CNI is a CNCF project and it is a curation of specifications and libraries. In the case of EKS, the CNI plugin uses overlay networks for communication and encapsulating the packets over the network. VPC CNI uses native VPC networking, which makes it highly scalable and results in high performance. It supports secondary **Classless Inter-Domain Routing (CIDR)** IP ranges, so you get more IPs from the secondary subnet. It is fully maintained and supported by AWS.

- **Storage**

  Stateful workloads are supported by Kubernetes through resources such as persistent volumes. Amazon EBS and **Elastic File Storage (EFS)** work with EKS through volume plugins. Depending on your workload requirements, you select EBS, EFS, or FSx to get the storage for your workloads.

The following diagram illustrates the process of deploying a cluster on EKS:

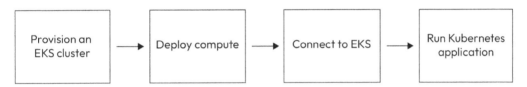

Figure 7.13 – EKS cluster deployment

The steps defined in the diagram are as follows:

1. Using Amazon EKS via the AWS console/CLI/**software development kit (SDK)**, you can provision a cluster.
2. All the required cloud resources are provisioned in the background.
3. Once the cluster is ready, you can use tools such as *kubectl* to communicate with your cluster.
4. You can start running your workloads once everything is connected to the Kubernetes cluster.

In the next section, let us take a look at the business use cases that will benefit from using Amazon EKS.

Use cases

Let us take a look at some of the application modernization use cases for EKS:

- **ML workloads**

  Kubeflow is an open source project to make deployments of ML on Kubernetes easy, portable, and scalable. AWS made a specific distribution of Kubeflow available to make the integration with AWS services seamless. EKS supports a wide variety of ML use cases, including computer vision, natural language processing, speech translation, and financial modeling, so you can train, build, tune, and deploy various ML models for these tasks.

- **Batch processing**

  You can run sequential or parallel batch workloads on your EKS clusters easily by using the Kubernetes Jobs API. Preparing, scheduling, and executing your batch computing workloads by leveraging the broad spectrum of AWS computing options and features of EKS makes it favorable to facilitate production batch workloads at scale.

- **Running web applications running Kubernetes on your own**

  If you are running web applications with a high operational burden of running Kubernetes on your own and do not have skilled teams, it becomes difficult to ensure your applications benefit from the performance, size, reliability, and availability that come from running on EKS using AWS.

- **Data on EKS (DoEKS)**

  You can accelerate your data journey and modernize your data platforms with Amazon EKS. The data on EKS can be broken down into six main use cases. The following diagram showcases all the open source data tools, Kubernetes operators, and frameworks that are supported by DoEKS. This also showcases the integration with the AWS Data Analytics managed services with DoEKS open source tools:

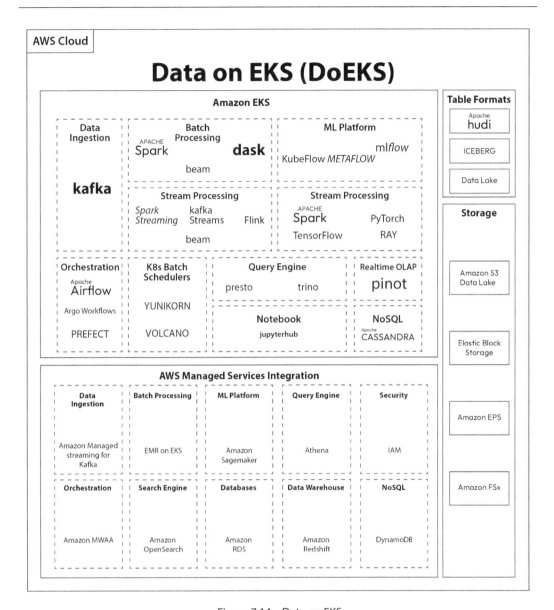

Figure 7.14 – Data on EKS

- **EMR on EKS**: You can submit Apache Spark jobs on EKS clusters and consolidate your analytical workloads on the same EKS cluster. This contributes to reducing costs, optimizing performance, and simplifying infrastructure management.

- **Data analytics on EKS**: If you have data analytics tools to extract insights from large and complex datasets, running those tools such as Apache Spark and Dask on EKS can greatly simplify scaling and rolling updates and ensure the high availability and reliability of your data analytic pipelines. Engineers and data scientists can focus on the analysis and interpretation of the data while EKS takes care of automating the deployment, scaling, and management of complex applications.

- **Artificial intelligence (AI)/ML on EKS**: If you are running AI/ML platforms using tools and technologies such as TensorFlow, PyTorch, and Kubeflow, you can make use of EKS to automate the deployment, scaling, and management of these complex applications.

- **Job schedulers on EKS**: If you are running data pipelines, ML workflows, or batch processing jobs, and use tools such as Apache Airflow, Argo Workflows, or Amazon **Managed Workflows for Apache Airflow (MWAA)**, then running them on Kubernetes can empower you by simplifying the automation and management of the job schedulers.

- **Streaming platforms on EKS**: If you are analyzing data in real-time or event-driven architectures and running streaming platforms such as Apache Kafka, Apache Flink, or Apache Pulsar, then EKS can help by providing benefits such as automatic scaling, rolling updates, and self-healing capabilities.

- **Distributed databases on EKS**: You can use EKS if you are using any popular distributed databases such as Apache Cassandra or Apache Presto.

ECS and EKS functionalities are quite similar in terms of allowing a mix of AWS compute platforms. In the next section, we will look at the differences between ECS and EKS to better understand the platform that you choose.

ECS versus EKS

With the options that AWS has to offer for container orchestration services, it becomes confusing which service to go with. While there are key differences between ECS and EKS, I would suggest analyzing these services and choosing the one that aligns with your application requirements and operational preferences. Let us first look at the differences between these services and then cover the key considerations to keep in mind while choosing between ECS and EKS.

Differences

Given that ECS and EKS are both fully managed container services, you will see a wide variety of compute options and deep integration with other AWS services. Let's break down the differences across the key aspects that would be essential for any enterprise:

- **Portability**

  ECS is an AWS homegrown technology where users are locked into the infrastructure and are not able to move the clusters to another cloud provider or on-premises as is.

EKS is a certified conformant Kubernetes service, where clusters can be easily migrated to another Kubernetes environment irrespective of the cloud provider. A cloud-agnostic platform is thus supported by EKS.

- **Simplicity versus flexibility**

    ECS is popular for the simplicity it offers. Running containers at scale comes inherent with ECS, and it drastically reduces the end users required to build, deploy, and migrate their containerized applications to this orchestration service. This opinionated service aims at reducing the number of decisions that you need to make around compute, networking, and security configurations. Features such as ALB and health checks integrate seamlessly with ECS, so you do not have to invest the time to build or operate any abstractions.

    EKS is popular for its ability to handle the heavy lifting of building and running workloads at scale. EKS offers access to the full capabilities of the Kubernetes features and functionalities at the convenience of the cloud provider taking care of operational overhead.

- **Security**

    ECS integrates with AWS IAM, so you get the standard level of control access that IAM offers.

    EKS requires add-ons to enable IAM functionally, and additional features such as Kiam come with additional costs and complexity.

- **Networking**

    At a high level, ECS provides networking options such as AWS VPC network mode, bridge mode, and host mode. These options work for most general use cases and have limited customization possibilities. EKS, on the other hand, provides fine-grained control over the network. You can customize how your Pod networks should work through custom CNI configuration. You can assign public and private addresses at the Pod level and even run on different subnets. These network mode options, however, are not available in EKS/Fargate.

- **Community support**

    ECS is not open source, and it does not have community contributions as such.

    EKS has a vibrant community and ecosystem where open source APIs offer broad flexibility. As of the 2020 CNCF Kubernetes Community Annual Report, there are 52,000 contributors, with 24 special interest groups generating close to 50,000 contributions. You get to leverage the wide ecosystem of Kubernetes tools, utilities, and extensions created thus far by the community.

- **Pricing**

  ECS comes at no extra charge for running clusters.

  EKS charges $0.10 per hour for each cluster as of 2023. This may add up to an additional $72 per month for every cluster you operate on AWS. This may add significant costs if you plan to operate multiple clusters. Keep in mind that this pricing only accounts for EKS control plane nodes. For the worker nodes, you will have to take the Amazon EC2 or AWS Fargate pricing, depending on the launch type you select.

	ECS	EKS
Portability	Low	High
Scalability	High	High
Simplicity	High	Low
Flexibility	Medium	High
Security	High	High
Networking	High	High
Community support	Low	High
Pricing	Low	High
Monitoring	Medium	High

The requirements of your applications and your preferences on how you want your teams to operate will guide your choices overall. Ensuring that you have consistent management tooling and take advantage of their simplicity will ensure you are modernizing your workloads on AWS seamlessly.

When to use ECS

Let us take a look at the following scenarios that are best suited to using ECS:

- ECS is well suited for workloads of any size due to its simplicity and feature integration set with the AWS ecosystem. It does not have a steep learning curve, so small enterprises or teams that are kick-starting their container journey do not have to invest in additional learning.

- Seamless AWS integrations allow end users to leverage resources such as ALB, Route 53, **Network Load Balancing** (**NLB**), and IAM to manage cloud-native architectures easily.

- Many enterprises use ECS as the stepping stone to EKS. If you want to gain experience with running containers at scale with low touchpoints, ECS can be your choice with a low upfront investment.

When to use EKS

On the other hand, EKS has its own shining areas with a full feature set and integrations suitable for the following scenarios:

- EKS offers unparalleled control over pod placement and resource sharing, making it well suited and invaluable for service-based architectures.

- The flexibility that comes with EKS empowers you to manage underlying resources with any compute platform, namely EC2, Fargate, on-premises via EKS Anywhere, and eventually migrate to any cloud provider without much hassle.

- EKS provides greater granularity when it comes to monitoring capabilities via Kubernetes built-in tools and available integrations via community support.

To sum it up, choose EKS if your teams are experienced with the Kubernetes platform and want to be cloud agnostic. Try ECS if you are just getting started with containers or are already running your workloads on AWS to leverage the AWS ecosystem.

Tools

There are multiple tools available in the form of services that you can use to build containerized applications on AWS.

AWS Elastic Beanstalk

Many organizations want to automate their CI/CD processes, such as building, testing, and deploying applications using continuous methodologies. With Elastic Beanstalk, you can manage infrastructure provisioning, monitoring, deployment, and packaging web applications into containers and achieve faster delivery cycles. This easy-to-use PaaS supports containerization on servers, including Nginx, Apache, IIS, and so on.

AWS Proton

Platform engineers can use AWS Proton and increase the rate at which they can define, and maintain deployments of applications. Proton can be used to connect and coordinate all the tools required for infrastructure provisioning, code deployments, monitoring, and updates.

This managed service provides infrastructure templates that can be used in a self-service model. Proton can be used to increase the velocity of the development and deployment process throughout an application life cycle. This service enables platform engineers to separate infrastructure as code into environment and service templates.

Developers can use Proton to deploy projects and provision infrastructure centrally without having to interact with the underlying resources. This helps in improving the developer's productivity and makes it easy to deploy their code using containers.

AWS App Runner

App Runner provides the simplest ways to build and run containerized web applications with a fully managed container-native experience. With this service, you don't need to worry about the following aspects:

- Servers to manage
- Configuring orchestrators
- Building pipelines
- Securing TLS certificate rotations
- Logging and monitoring
- Configuring VPCs
- Auto Scaling

The complexity is abstracted away from the preceding tasks, and developers can run containers on AWS without even managing any operational considerations.

Amazon A2C

A2C is a command-line tool that can be used to migrate Java and .NET web applications into container images. Developers can analyze and build applications running on infrastructure such as bare metal, VM, or EC2 instances in the cloud. Developers can use A2C to package the application artifacts and their dependencies into deployment artifacts that can be deployed and run using Amazon ECS or EKS.

All the required infrastructure, including configuring network ports and CI/CD pipelines on AWS, is provisioned by A2C. These modernized container applications can be enabled by leveraging the best practices that AWS incorporated into these services without any code changes.

AWS Copilot

Copilot is an end-to-end developer workflow tool that offers a CLI to deploy and operate containerized applications on Amazon ECS specifically. Constructs such as ALBs, public subnets, tasks, services, and clusters are configured and created on your behalf. Copilot lets you deploy across multiple environments, accounts, and/or Regions through the AWS CodePipeline.

ROSA

Enterprises that have been running applications using Red Hat OpenShift on AWS can benefit from ROSA to reduce effort and meet agility requirements easily. ROSA helps such enterprises accelerate customer deployment and migration to the cloud by dramatically reducing the efforts of purchasing and procuring Red Hat OpenShift subscriptions, which is often time-consuming and can be a challenge to do when running at scale.

ROSA provides a consumption-based model where OpenShift end users can spin clusters up and down on a need basis rather than a subscription basis. Developers do not have to deal with infrastructure as code templates or automation scripts to deploy OpenShift.

EKS Anywhere and ECS Anywhere

These two services, which were recently launched in 2021, provide you with the option to run your on-premises environment on AWS. This caters to those enterprises that want to run workloads on-premises due to data residency, latency, regulatory, or compliance requirements but still want to take advantage of the cloud benefits. In order to address such scenarios, ECS Anywhere and EKS Anywhere let you run workloads on AWS Outposts in your own on-premises facilities.

This particularly helps enterprises that have already invested in their own IT infrastructure but would like to easily integrate with management and deployment mechanisms that are available on AWS. You can choose between ECS Anywhere and EKS Anywhere for such workloads.

Case study for Amazon ECS

The case study in this section covers a customer's journey of using Amazon ECS.

The business challenge

EducatEra is an educational technology company considered to have built an innovative platform that provides universal access to the best educational and interactive courses. The company partners with various top universities and organizations to offer online courses worldwide. With more than 15 million users across 150 countries, EducatEra has more than 2,000 courses to offer from 120 institutions. EducatEra had a large monolithic application and considered migrating from its on-premises infrastructure. A careful decision had to be made given that EducatEra is a risk-conscious company. They wanted to build an innovative platform and enhance their current application to support advancement in the education space and anticipate future business needs. They also wanted to implement DevOps practices and produce new courses at a faster rate. The company had to cut down costs and get rid of its IT departments that managed servers and operational activities. They had the following additional challenges:

- The monolithic application had performance issues.
- The tightly coupled architecture of the monolith increased the blast radius during system failures.
- They had dependencies on third-party licensed applications.
- There was increased downtime due to the maintenance cycles.
- Manual deployments were required for new feature releases.

All the preceding challenges made the company explore the modernization path for their monolithic application and migrate to AWS using a microservices architecture.

Solution overview – Amazon ECS

EducatEra performed application discovery and revamped its monolith application to make improvements to its existing architecture. The development team used the famous Strangler pattern to decouple the new architecture into various microservices, such as course catalog, subscription service, customer account service, search engine, and course playlist. The microservices were containerized using Amazon ECS, which can run Docker containers without having to manage any servers or clusters. Amazon ECS performs the container orchestration, and the Docker container images were stored in Amazon ECR.

To set up the CI/CD pipeline, EducatEra used Jenkins for CI/CD. Running Jenkins in Docker containers allowed them to make efficient use of pipelines for automation and remove the maintenance aspect during deployment and releases. The independent deployment of microservices was automated, and this maximized the business benefit, thereby improving granular compute resource allocation and greater performance, availability, and reliability. EducatEra successfully migrated 70 plus microservices after the deployment of microservices on AWS and achieved significant performance improvement by using load balancers and caching techniques.

Case study for Amazon EKS

The case study in this section covers a customer's journey of using Amazon EKS.

The business challenge

A business chat tool company chat4business moved to a container-based architecture a few years ago. The company's infra team was self-managing Kubernetes on Amazon EC2. However, the administrative and operational load of performing version upgrades every three months was an overhead to the company. There were issues with the tooling at the onset of Kubernetes adoption and the operational load to manage each cluster was high. The company had been using open source tools to manage its Kubernetes clusters. The infra team had to do everything including building, managing, and operating the control plane themselves. As their systems grew in scope and complexity, the operational load increased. The company needed failover risk during deployments and wanted to automate operations wherever possible in order to avoid dependency on any particular individual.

Adopting Amazon EKS

For all the reasons mentioned in the previous section, chat4business migrated to Amazon EKS, with an SLA-based phased release strategy. Migrating its messaging application platform and its PHP web application platform to Amazon EKS took four months. The time to perform failovers when release issues occur was cut down by 95%, and overall operational workloads during releases dropped by 90%.

Offloading the management of the Kubernetes control plane helped the infra team focus on improving applications with declarative Kubernetes deployments. The team was able to automate software delivery during cluster updates. The team was also able to focus on strengthening security through penetration tests, security scans, and introducing security visualization tools for container environments. Additionally, the team was also able to optimize costs of up to 80% as a result of adopting and integrating Amazon EC2 Spot Instances into its infrastructure.

Summary

In this chapter, we discussed in detail the AWS compute services that are available to take advantage of and implement application modernization for your workloads. Various use cases were covered for each of these services, and in-detail comparisons were presented to enable you to make the right decisions while choosing the services for your modernization efforts.

In the next chapter, we will continue with integration and developer agility-specific services that you can leverage to streamline IT footprint modernization using AWS.

Implementing Compute and Integration on the Cloud Using AWS

As the cloud continues to be the top enabler of your business outcomes, modernizing infrastructure and integration paths is an integral aspect of digital transformation, among other business developments. The primary reason leaders value digital modernization is growth and stability. Aligning with digital transformation enables you to drive growth, processes, people, and culture. As business leaders, acknowledging the necessity and power of digital transformation is important to facilitate innovation and continually evolve.

There are many barriers when it comes to digital modernization: legacy systems, operating under heavy regulatory scrutiny, and unfamiliarity with the potential of cloud technologies to deliver on business objectives, to name a few. In this chapter, we will focus on what AWS has to offer and take steps toward cloud-enabled changes. With AWS' offerings and from its substantial evolution, leaders like you can make notable progress by pursuing steps such as adopting the right mental model, capitalizing on the cloud as an enabler, and aligning on modernization.

We will cover the following topics in this chapter:

- Introduction to serverless
- Serverless computing on AWS
- Containers versus serverless on AWS
- Case study on serverless
- Introduction to application integration services and iPaaS

- Diving deep into API management, Event Bus, and messaging on AWS

- Case study on AWS integration services

- Introduction to AWS ALM services

- Diving deep into AWS IaC tools

- Case study on AWS IaC tools

Introduction to serverless

Serverless architecture and technologies on the cloud are becoming more and more relevant for modern applications. Today, we are seeing businesses making a radical shift toward serverless as cloud providers offer serverless products. In this section, we will explain what serverless is, dive into some of the benefits of using serverless, and explore some of the common use cases that are a good fit for serverless.

What is serverless?

Serverless can be best described as an execution model that allows developers to build applications and focus on just that. There is no requirement for additional infrastructure or operations teams to maintain the servers or dependent software.

Benefits of serverless

If you are a developer, manager, or business stakeholder, consider the following benefits as to why you should consider making serverless a priority:

- **Move from idea to market faster**

   Many IT firms hire developers not just to write code but also to build a product that is financially viable and has a positive impact on their customers. To deliver a product idea, there is a lot that happens behind the scenes apart from implementing the functionality itself. The following diagram depicts the milestone steps that are involved, from ideation to production launch:

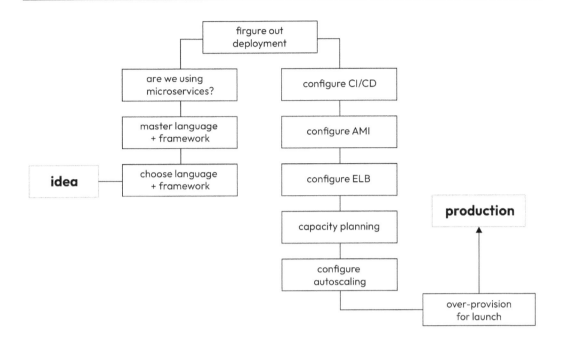

Figure 8.1 – Ideation to production

With serverless, many of the steps shown in the preceding diagram are dramatically simplified. You get the freedom of writing code, running the applications, managing the data, and integrating applications without having to manage servers and infrastructure. Developers get to focus on implementing product features, which is a more productive use of their time and energy. Tasks involved in managing the infrastructure, such as server provisioning, security patches, and scaling, don't have to be handled.

- **Lower your costs**

    Cloud providers charge for what you use with serverless. Requests are API-based and you get the granularity to see all the usage and billing details based on how many requests were made or how many seconds a particular function took to execute. You don't have to predict how many resources you will need at any point in time or end up over provisioning infrastructure to accommodate the increase in demand. Administration overhead is removed, which is an additional cost saving right there.

- **Adapt at scale**

    Serverless design patterns in the cloud allow you to focus on innovation and achieve throughputs at any given point in time. The ability to respond quickly to demand is the most important aspect for businesses to succeed in today's digital world. Serverless computing enables your systems to scale and eliminate the tremendous overhead of managing servers.

Automatic scaling, adapting to the rate of incoming requests as they come in, and zero impact on application performance makes serverless computing platforms extremely desirable for applications that are mission critical.

- **Build better applications, more easily**

  Developers building applications on self-managed, server-based computing platforms face challenges such as timeouts, throttling, downstream dependency slowdowns, and increased latencies. As the load increases on systems, the infrastructure comes under stress and identifying points of failure in the systems becomes challenging. Serverless computing platforms offer built-in capabilities that enable you to focus on building your applications instead of configuring desirable state.

Common use cases for serverless applications

If you are running workloads that are asynchronous and concurrent and have sporadic demand and unpredictable demand in scaling requirements, then serverless can be a good option. Let us take a look at some of the top use cases that best fit serverless:

- **Multimedia processing**: Scalable event-driven processing and the elastic capabilities that serverless computing offers are favorable to achieve increased execution times and the ability to use GPU-based resources. Image and video manipulation at scale can be achieved with serverless.

- **Polyglot or multi-language applications**: Applications that are built using different programming languages can thrive easily using serverless for the seamless integration of services. With serverless, applications can be multilingual, and it encourages teams to explore and choose the language that best fits their application needs rather than getting locked into using the same language.

- **Web applications and SaaS**: Modern cloud applications are increasingly modular, heavily utilizing third-party **Software as a Service** (**SaaS**) software to deliver functionality. Serverless opens rich possibilities to handle HTTP REST APIs and lets you subscribe APIs to third-party events, regardless of where they are hosted.

- **Batch jobs**: Many organizations process large amounts of data in the form of batch jobs. Examples include **billing**, **report generation**, **data format conversion**, and **image processing**. Serverless enhances the overall performance of batch processing applications through intense parallel computation, I/O, and network access.

- **Continuous Integration and Continuous Deployment (CI/CD) pipelines**: Serverless provides the ability to automate CI/CD pipelines easily. This helps in rapidly iterating software where features or bug fixes can be shipped at a much faster pace.

In summary, serverless computing gives the developer community the ability to focus solely on coding and developing their applications. With the simplicity and cost savings that serverless computing can drive, it makes it easy to auto-scale horizontally without doing capacity planning or ongoing maintenance.

Serverless computing on AWS

Serverless on AWS is a technology to build and run applications using AWS services as building blocks and not think about managing the underlying infrastructure. AWS offers computing power to run your workloads at scale to meet the real-time demand of your businesses. AWS is responsible for handling all backend tasks such as computing, data storage, processing, and architecting for high availability.

There are serverless offerings across the three layers of the stack of a modern application: **Compute**, **Integration**, and **Data Stores**. We will go through compute and integration in this chapter.

In serverless computing, we have two main services: **AWS Lambda** and **AWS Fargate**.

AWS Lambda

When building serverless applications on AWS, AWS Lambda is one of the core services for running application code. AWS Lambda is a serverless computing service that is event-driven, with which you can create and run self-contained applications. You can run code (supported platforms are Java, Go, PowerShell, Node.js, C#, Python, and Ruby) with high availability and be free of administrative efforts such as operating system maintenance, capacity provisioning, code, and security patch development.

Typical web applications need a computing service, a database service, and an HTTP gateway to support the application stack. Lambda's seamless integration with many other AWS services, such as API Gateway, DynamoDB, and **Relational Database Service** (RDS), makes it a great fit for a wide range of applications. Its rich ecosystem allows developers to build serverless applications that can be easily deployed and published through the AWS **Serverless Application Model** (SAM).

How does Lambda work?

AWS Lambda takes your code and runs it in its own container. Say, for example, you create a function such as the following one; Lambda packages it into a new container, and executes it on a multi-tenant cluster of servers managed by AWS. A function has code such as this:

```
exports.handler = async (event) => {
  return {
    statusCode: 200,
    body: JSON.stringify({ msg: "Hello from Lambda!" })
  };
};
```

Each Lambda function is allocated a specific amount of RAM and CPU capacity and you get charged based on the allocations and the amount of runtime the function takes to complete its execution. Lambda is powered by a virtualization technology called **Firecracker**. Lambda uses Firecracker for rapid provisioning with a minimal footprint and for running functions on secure clusters.

> **Firecracker**
>
> Announced in 2018, Firecracker is a new virtualization and open source technology to allow secure multi-tenant, container-based services to run at a much-improved speed with resource efficiency and improved performance.

Lambda functions do not run constantly and are triggered by events, which is called **invocation**. All interactions with the code are made through the Lambda API and Lambda acts as a glue between the various AWS services in your architecture and the business logic. This makes it ideal for Lambda to fit in well with event-driven architectures.

Features of Lambda

Here are some of the features of AWS Lambda:

- Bring your own code is supported so you can package any code as a Lambda layer and it can be reused across multiple functions. Java, Go, PowerShell, Node.js, C#, Python, and Ruby code are natively supported, and any additional programming languages are supported by a runtime API.

- Automated administration helps add to the operational simplicity of Lambda where sub-second startup times, automatic scaling, high availability, and native integration with over 200 AWS services and SaaS applications makes it a highly favorable service.

- AWS Lambda supports packaging and deploying functions as container images, making it easy for developers to use container image tooling, workflows, and dependencies that they are already familiar with.

- You can connect to databases seamlessly to take advantage of relational databases such as MySQL and Aurora through RDS Proxy. Amazon DynamoDB Streams is integrated with AWS Lambda to react to data modifications in a DynamoDB table. Through triggers, Lambda functions are invoked synchronously to respond to events in DynamoDB Streams. For example, you can configure a trigger when an item in a table is modified to send a notification via Amazon SNS.

- Flexible integration enables Lambda to seamlessly plug into your favorite monitoring, observability, security, and governance tools. With this, you get to capture fine-grained telemetry and manage logs, metrics, and traces effectively.

- Powered by Graviton2, it can help you achieve at least 30 percent better price performance and can be useful for a variety of serverless workloads such as web applications and data and media processing. With better performance, low costs, and the highest power efficiency, it is the most favorable option for mission-critical serverless applications.

As a developer, you can focus on building the core business logic of your applications and not worry about managing operating system patching, right-sizing, provisioning, scaling, and other administrative tasks.

Use cases of Lambda

In addition to the serverless use cases that we discussed in the *Common use cases for serverless applications* section, here are a few powerful ways that you can leverage Lambda to realize the full value of running modern applications on the cloud:

- **Real-time data processing applications**: Lambda can be used to execute real-time data processing applications such as processing transaction orders, analyzing click streams, cleaning data, generating metrics, filtering logs, analyzing social media data, and performing IoT device data telemetry.

- **Real-time notifications**: Lambda can be used to monitor your cloud resources and detect security breaches. An alert can be triggered on specific configured events based on the AWS activity logs that get stored in AWS CloudWatch and CloudTrail.

- **Predictive page rendering**: Lambda can be used for predictive page rendering applications to retrieve documents and multimedia files and process the display page rendering.

- **Automated file synchronization jobs**: You can make use of Lambda by invoking it via scheduled events via Amazon EventBridge and you only pay for the execution time.

- **Chatbots**: Lambda can be used to build a scalable chatbot where all the commands are invoked via API Gateway and event-driven architecture can trigger the bot when there are specific API requests.

- **Change data capture**: Lambda can be used to trace changes in database accounts and keep track of changes made within databases such as Amazon DynamoDB. The seamless integration of Lambda and DynamoDB makes it favorable for you to leverage services such as DynamoDB Streams and notify your Amazon SNS function to send updates to subscribers whenever new transactions occur in the database.

Additionally, there is a wide range of other use cases, such as ETL processes, executing server-side backend logic, and creating web applications. Also, for functions that act as orchestrators, using AWS Step Functions will give you optimal results, rather than using AWS Lambda, when it comes to performance. Lambda functions that transport data from one service to another can often be replaced by service integrations.

Things to consider while using AWS Lambda

While you get to avoid infrastructure management, there are some limitations that you need to be aware of:

- **Payload restrictions**: Invocation payloads for Lambda have a limit of 6 MB, so you need to construct your HTTP requests accordingly. Upload any files to Amazon S3 instead of managing them in the application directly.

- **Invocation times**: Lambda invocation times have a limit of 15 minutes of runtime per function execution. If you have longer tasks, consider breaking them up into multiple Lambdas. This is where it is important to do a cost analysis and consider running containers.

- **Concurrency**: The default quota of concurrent executions is 1,000, although you can request up to tens of thousands. If you are running your Lambda at scale, the concurrent execution limit can potentially be a roadblock. It is important to consider employing a **content distribution network** (**CDN**) depending on the traffic analysis or even spreading your Lambda across multiple regions or AWS accounts.

- **Cold starts and latency**: Lambda functions have the concept of cold start. Let us look at what cold start means and its relation to Lambda invocations.

When a request is sent via a function to the Lambda API, the service needs to initiate the preparation of an execution environment. Lambda performs a series of tasks during this step:

1. Downloading the function code

2. Storing it in an internal Amazon S3 bucket or Amazon ECR if the function uses container packaging

3. Creating an environment with the specified memory, runtime, and configurations

4. Running any initialization code outside of the event handler before running the handler code

    The following diagram depicts the two steps of setting up the environment and code:

Figure 8.2 – Cold starts and warm starts

This is referred to as *cold start*. Even though you are not charged for this time, the overall invocation duration is impacted by latency.

Lambda retains the execution environment for a period of time if another request is invoked for the same function and the service can reuse the execution environment. This helps in improving resource management and performance and is referred to as *warm start*.

In order to reduce cold starts, *provisioned concurrency* is recommended where the function is initialized and ready to respond without any latency. Provisioned concurrency is designed to make functions available with double-digit millisecond response times, so interactive workloads will benefit the most from this feature.

AWS Fargate

A serverless compute offering for containers, AWS **Fargate** lets you run containers without managing infrastructure. Fargate is technology compatible with both container orchestration solutions: Amazon ECS and Amazon EKS. Fargate can also be used with AWS Batch application orchestration to run batch workloads in the cloud. The undifferentiated heavy lifting of container management is removed with Fargate and so developers can leverage the speed, agility, and immutability of the container's deployment method.

Fargate with ECS

With Fargate as the launch type, you can run containers without having to own, run, and manage the lifecycle of a compute infrastructure. The traditional container orchestration capabilities are all covered by Fargate technology. In the case of ECS, APIs such as `CreateService`, `RunTask`, and `UpdateService` are taken care of for you by Fargate and you don't have to worry about the EC2 instances that you typically would while running ECS workloads. You get billed based on the number of CPU cores and the amount of memory your ECS tasks consume, per second, and not for the EC2 capacity that goes unused if provisioned through traditional ECS with the EC2 launch type. The following diagram is a visual representation of how Fargate as the underlying technology takes care of the essential orchestration operations.

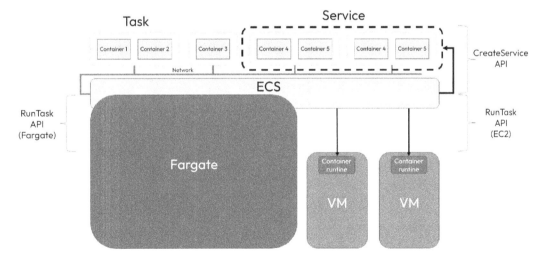

Figure 8.3 – Fargate with ECS

Running containers with a Fargate-managed fleet gives you the additional data plane option of running your tasks parallel to EC2-based clusters.

Use cases

Running your containers with Fargate is a great option for the following scenarios:

- Workloads with low overhead as a priority: Fargate takes infrastructure management out of the equation for you to focus on just the business. Managing a large cluster of EC2 instances where you have to patch, secure, and maintain the latest version of Docker and the ECS agent incurs costs and uses resources. Fargate automatically takes care of these behind the scenes, and you don't have to worry about the underlying infrastructure.

- Workloads with real-world scaling: Fargate works well with web applications that are containerized and have expected traffic. Fargate comes with 16x faster scaling, improved performance, and reduced wait times as of 2022, to run applications at a large scale. You can burst up to 100 on-demand or spot tasks with a task launch rate of 20 tasks per second.

- Batch workloads: Fargate is an excellent choice for batch jobs with up to 1,200 on-demand tasks. You can launch your batch jobs in less than a minute. Cron jobs or jobs that come from a queue are a perfect fit for Fargate. If you have batch workloads, it is recommended to use Docker images with AWS Batch for cost-effective and fast processing of complex jobs. Queueing workloads, prioritizing jobs, handling dependencies and retries, and scaling compute are all simplified using AWS Batch and it also helps with lifecycle management.

Fargate and Lambda are comparable when it comes to being part of the serverless technologies that AWS has to offer. There are key differences with respect to the CPU/RAM available for each service to use. A Lambda function can use a maximum of 3 GB of memory as of 2022. Fargate has more flexibility, where you can go up to 30 GB of RAM and 4 vCPUs as of 2022.

Limitations

There are a few limitations with Fargate and ECS that you need to consider before choosing this technology for your workloads:

- With fewer customizations and less fine-grained control, if you require greater control of your EC2 instances to support compliance and governance requirements, then you may need to consider ECS without Fargate. Runtime access for users such as SSH or interactive Docker is not supported with Fargate.

- Fargate does not support task definition parameters that are relevant to hosts such as `gpu`, `ipcMode`, and `extraHosts`.

- Fargate tasks only support mounting host volumes so `dockerVolumeConfiguration` is not supported.

Fargate with EKS

Fargate's integration and support for EKS were introduced in 2019, with Fargate acting as a vending machine and becoming a container data plane for Kubernetes. Without having to provision or configure virtual machines, you get the on-demand, right-sized compute capacity for your container workloads. The following is a high-level visual representation of how Fargate on EKS works.

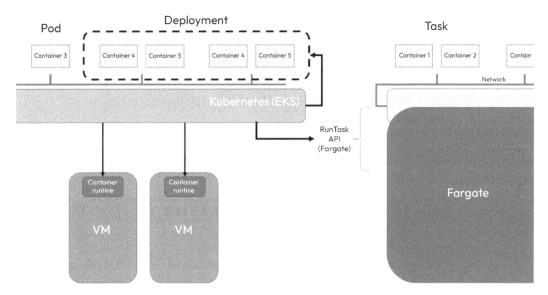

Figure 8.4 – Fargate with EKS

The custom scheduler running on EKS is decoupled from Fargate and the Fargate scheduler invokes the `RunTask` API to schedule pods. Fargate uses the AWS controllers to integrate with EKS and upstream Kubernetes. These controllers are part of the Amazon EKS managed control plane and are responsible for the native pod scheduling in Fargate.

Limitations

There are a few limitations with Fargate for EKS that you need to consider before choosing this technology for your workloads:

- There is no support for daemon sets, so if your application requires a daemon, then you have to consider configuring a sidecar container in your pods.

- The region availability of EKS with Fargate is still an evolving aspect – as of 2022, EKS with Fargate is not supported in the China and GovCloud regions.

- Startups are slow relative to an EC2 instance that is already connected to and known by the cluster. But there are other dimensions, such as the hidden costs of operations and the value of AWS Fargate, that can lead you to opt to use a managed service such as Fargate.

NLB and ALB support exists only for Fargate with IP targets. Additionally, Fargate only supports ECR or Docker Hub, which can be a constraint if you are storing your container images in private registries. When running containers using Fargate, there are service limits, such as the number of public IPs, the maximum container storage size, and so on, which you need to request increases for.

Fargate is gaining more popularity and appeals to many businesses, with sysadmin skills not needed to maintain the infrastructure supporting containers. Whether you are just at the beginning stages of evaluating container approaches or established containers are the de facto standard for your organization, Fargate certainly helps you in realizing the operational and economic benefits of containers.

Compare Fargate services

With all the orchestrator options we have discussed, it is natural to get overwhelmed, so here is a simple cheat sheet that you can take a quick look at and identify the differences easily:

	ECS + Fargate	EKS + Fargate	ECS + EC2	EKS + EC2
Deployment	AWS only	AWS only	AWS unless using ECS anywhere	Any cloud provider supporting Kubernetes or on-prem
Cluster Costs	No cost	$$ Check AWS pricing	No cost	$$ Check AWS pricing
Compute Costs	$$ Check AWS pricing	$$ Check AWS pricing	$$ Check AWS pricing	$$ Check AWS pricing
Pricing Options	On-demand instances, Spot instances, Savings Plan	On-demand instances, Savings Plan	On-demand instances , Spot instances, Savings Plan	On-demand instances, Spot instances, Savings Plan
Virtual Machine	No need for EC2 management	No need for EC2 management	EC2 instances on your AWS account	EC2 instances on your AWS account
Infrastructure scaling	AWS	AWS	You need to configure ECS Cluster Autoscaler	You need to configure Kubernetes Cluster Autoscaler

Networking	ENI and security group per task/container	ENI and security group per pod/container	ENI and security group per task/container	Multiple pods share the same ENI and security group
Load Balancing	ALB, NLB	ALB, NLB	ALB, NLB, CLB	ALB, NLB, CLB
Service Discovery	Cloud Map	Cloud Map	Cloud Map	Cloud Map
Use Cases	Workloads with low overhead as a priority, occasional bursts, batch workloads	Web applications, APIs, and microservice architectures that require lifecycle management, workloads with AI and ML development and the need to boost server capacity without overprovisioning	Workloads that have requirements such as running machine learning tasks on GPU instances, persistent storage with EBS volumes, or more control of networking	Workloads in hybrid environments, distributed training jobs using GPU-powered instances, and deep training inferences using Kubeflow

Table 8.1 – Comparison chart

In the next section, we will look into the differences between containers and serverless and get an understanding of the scenarios in which each of these technologies best fits your applications.

Containers and serverless computing on AWS

Containers and serverless computing are technologies that enable you to build applications and ease the management overhead of infrastructure. There are similarities in the nature of workloads that both containers and serverless can support:

- Stateless workloads
- The ability to abstract away the underlying host environment from applications
- Dynamically scaling workloads
- Simplifying the overhead and heavy lifting of managing virtual machines

There are also some striking differences. In this section, we will discuss some of the aspects that you will need to analyze before choosing one option:

	Containers	Serverless
Deployment Options	Support Linux and Windows	Specific hosting platforms that are compatible with other cloud providers' equivalents to AWS Lambda
Cost	Pay for the AWS resources used, for example, EC2 for ECS and EBS if you attach storage volumes.	Pay-per-use
Virtual Servers	You manage your application infrastructure and ensure the servers are patched regularly.	AWS takes care of servers and any underlying infrastructure.
Scalability	You manage the scaling of the server capacity that is required to run your containers.	The infrastructure is scaled up or down automatically to meet the demand requirements.

Table 8.2 – Containers versus serverless

Which technology you choose will depend on the requirements of your application, your use case, and your team's expertise. In a nutshell, here are the benefits and use cases of each technology:

Containers	Serverless
Benefits: Containers are ideal for complex applications that can benefit from a high level of control of the operating system and with longer execution times, speeding up processing times to run workloads in a containerized environment.	Benefits: Serverless is ideal for applications that require an expansive number of releases for features and functionalities. There's a reduced possibility of errors, and event-driven architectures will favor serverless.
Example: A large e-commerce website comprising several functionalities, with features such as pricing, payment options, inventory management, product catalog, mobile-friendly interfaces, and user reviews. These features can be designed in a modular fashion where each microservice can be packaged in a container.	Example: IoT applications such as smart home security systems, connected appliances, and wireless inventory trackers using event-driven architectures. The functions will be invoked only when specific events occur and it's strictly a pay-per-use model, allowing cost optimization, especially for cases with an unpredictable load.

Table 8.3 – Benefits of containers and serverless

As noted earlier, the design of the application and the infrastructure to support it will be key aspects to determine the form of technology that will best fit. Containers and Serverless technologies are paving the way for radical IT transformation and are redefining paths to modernization and how we perceive infrastructure, deployment models, and even business models.

Serverless and containers together!

With Fargate technology, AWS combines the relative advantages of serverless and containers to compensate for the weaknesses of each. Whether to use containers or serverless products, is a decision that you have to make considering factors such as cost, scaling properties, and the operational control you want to have. We have already discussed Fargate in detail, in the *AWS Fargate* section, and how it fits into the modernization picture.

Case study on serverless

In this section, we will do an in-depth analysis of how serverless technologies can enable businesses to make the choice to shift their existing applications running on-premises or to the cloud. The customer names are imaginary, but the use cases are real. These case studies are being shared with you so you can use pieces of the case studies and apply them as needed in your organizations.

AWS Lambda

AWS Lambda and other AWS cloud services such as API Gateway can be greatly beneficial for serverless computation. The following case study highlights one such usage pattern that is cost-efficient and requires less management.

Business case/challenge

A SaaS company where traffic peaks during Christmas gets an increased number of tickets where users are reporting 503 errors upon invoking certain public-facing APIs. These failures are impacting users on the biggest sales day of the year. Upon running several tests, the IT team finds out that their current systems cannot support the highest throughput.

Modernizing applications using AWS Lambda

In order to meet the demand, a new tech stack to modernize the current architecture was proposed and implemented. The new stack had to be economically viable and have endpoints in multiple AWS regions to provide low latency to all users across the globe.

This is a common pain point for many customers who like to run their services at a global scale and deploy their backends to multiple regions.

A new stack incorporating a serverless API in multiple regions leveraging Amazon Route 53 to route the traffic between regions is implemented as shown in the following architecture diagram:

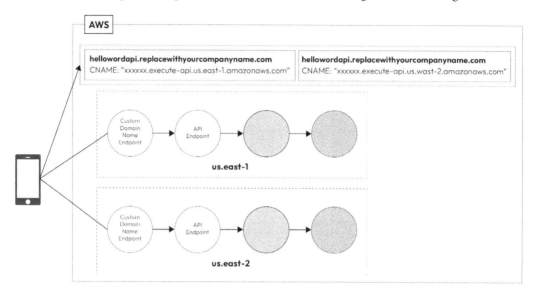

Figure 8.5 – New stack with multi-regional API Gateway

By refactoring the legacy application into AWS Lambda functions and using API Gateway to serve the API requests, the company is able to serve 25 million API calls per day. AWS Lambda enables the creation of functions in milliseconds, and as a result, the AWS bill is generated for what you use. With Amazon API Gateway, developers can deploy applications to AWS and can communicate with greater efficiency.

AWS Fargate with ECS

Customers of all sizes and any vertical can leverage AWS Fargate to help automate repeated processes and accelerate their project setup. The following hypothetical case study covers a scenario on how enterprises that have stringent regulations can still modernize their workloads on the cloud using AWS Fargate.

Business case/challenge

A large enterprise with heavily regulated protocols belonging to the financial services industry decided to migrate from on-premises infrastructure to AWS. This enterprise is risk-conscious and had to be comfortable with the value proposition before starting the modernization path on the cloud.

Modernizing applications using AWS Fargate

To keep up with advancements in the technology space and meet their future business needs, the company migrated 1,000 legacy applications to a microservices environment that were approximately implemented by 3,000 developers. The company implemented DevOps practices and produced microservices at a faster rate.

In order to support the scope of the scalability, cost efficiency, and disaster recovery capabilities, they chose AWS Fargate with ECS. The internal teams were able to quickly improve security through application isolation by design using Fargate. The company was able to get much more stability as it didn't have to provision infrastructure and as a result, reduced labor and hardware costs.

In order to implement the architecture shown in *Figure 8.6*, the company executed the following steps:

- Containerized the core Java application
- Configured the reverse-proxy server
- Containerized the NGINX reverse-proxy server
- Created the `docker-compose` file that will be used to deploy the services to ECS
- Pushed container images to Amazon ECR
- Created the ECS cluster and the ECS profile
- Created an application load balancer
- Created an AWS Fargate task definition
- Created the Amazon ECS service

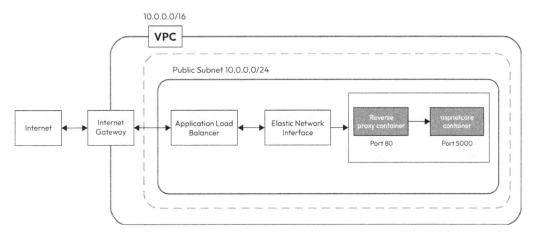

Figure 8.6 – Reference architecture for deploying to AWS Fargate

Additionally, the company was able to seamlessly implement the system based on a consumption model with increased deployment frequency, thereby accelerating the migration path. The company adopted multi-region strategies for its critical workloads and increased resiliency.

Introduction to application integration services and iPaaS

Irrespective of the technologies you choose, integration is an essential aspect that will shape and scope your modernization efforts across organizations and industries. Application integration platforms provide you with a way to enable your applications, each built for a specific purpose, to work with each other. By helping bridge the gap between these applications, you can support agile business operations and modernize your integration pathways to operate more effectively and efficiently.

APIs are a common design pattern to modernize applications and can enable cloud-native applications to securely expose their capabilities. **Integration Platform as a Service** (**iPaaS**) helps to provide a single platform for organizations that have a wide variety of applications. These are typically very difficult to manage, and it becomes a hassle for IT teams to coordinate processes and workflows.

> **Integration Platform as a Service (iPaaS)**
>
> iPaaS is a platform as a service that provides a set of services to standardize the integration of different types of applications and enable business process automation at improved speed and accuracy across an organization.

There is a rapid shift toward iPaaS to standardize integration capabilities and unlock new opportunities. Businesses want to modernize on the cloud and reduce the time taken to build the integration capabilities of their systems.

The challenge that iPaaS solves

As per market research done by **chiefmartech** (`https://chiefmartec.com/wp-content/uploads/2022/05/state-of-martech-2022-report.pdf`), there were 150 business applications in 2011 versus 7,000 business applications in 2019 alone for integration services. There has been a SaaS sprawl in the market today when it comes to the number of SaaS applications available to tackle specific integration tasks and address common enterprise integration challenges. Teams in the same organization end up using siloed services that fit with the workings of the products they build and do not always think of the bigger picture of how all of these would play across multiple departments.

As the volume of applications grows, so do the silos, leading to manual management of all these services, and increasing the lack of visibility and costly subscriptions. The concept of integration is not new, but it has always been an afterthought for integration projects to be initiated on an as-needed basis. iPaaS enables organizations with a typically cloud-based integration platform for applications to connect easily.

AWS application integration services provide a suite of integration capabilities in the following five categories:

- **API management**: Consisting of tools and services that developers can leverage to build, analyze, operate, and scale APIs in the products that they build.

- **Event-based communication**: To connect your application to your own apps, SaaS, and other products.

- **Messaging and message queues**: To support reliable platforms that can enable communication between different systems and platforms. When it comes to modern cloud architecture, applications are mostly modular for the reasons we discussed earlier. These applications communicate with transportation channels such as point-to-point requests, message queues, topics, and event buses.

- **No-code API integration**: To connect different SaaS applications and AWS services at any scale without writing code.

- **Service orchestrators**: To manage distributed applications that are heavily dependent on external services in workflows so you can automate IT and business processes easily.

As an enterprise leader, you have the opportunity to take iPaaS platforms to the next level and embrace the change in contrast to building traditional out-of-the-box integration solutions.

Benefits of application integration

The concept of application integration and its adoption is applicable for businesses of any size for the following reasons:

- **Increased workplace efficiency**: As a result of the simplification of communication and reducing the time and effort of manual setup.

- **Reduced cost of business operations**: You don't have to maintain multiple applications for your business operations and end up being impacted by increased costs for tools usage and maintenance.

- **Improved workflow automation**: Bringing all software solutions used by a company under one roof helps you achieve better control over them. Better information management ensures leaders and managers make informed decisions that suit their business the most.

- **Improved interoperability**: For large organizations where interoperability is one of the biggest challenges, an innovative app integration solution will typically ensure interfaces are provided to streamline multiple business operations.

- **Increased innovation**: Presenting a chance for employees to identify opportunities that better fit your organization.

To fully integrate and connect your applications, it is necessary for you to look at suitable application integration technologies and see what best fits your organization's work processes. In order to that, we will cover the AWS offerings in the iPaaS space in the next section.

Diving deep into API management, Event Bus, and messaging on AWS

When it comes to AWS application integration services, there are multiple offerings that are available on AWS. We will not cover the breadth of all the services, but we will highlight a few services and look in detail at how they benefit real-time application integration processes.

Amazon API Gateway

The application-facing component of a typical AWS serverless infrastructure, API Gateway provides capabilities to create, maintain, monitor, publish, and secure APIs based on REST, HTTP, and WebSocket.

Features

Here are some of the features that make API Gateway favorable for modern applications:

- It supports canary deployments to safely roll out any changes to your APIs.
- It supports both stateful and stateless APIs.
- Logging and monitoring are integrated with AWS CloudWatch and AWS CloudTrail.
- Authentication is integrated with AWS IAM and Amazon Cognito user pools.

API Gateway offers scalability that is ideal for containerized and serverless workloads. With the ability to handle thousands of concurrent API calls and perform traffic management, CORS support, along with throttling and monitoring, API Gateway can be used to optimize serverless workloads.

Use cases

Three distinct use cases of API Gateway are as follows:

- **Serverless applications** require a unified interface and a connection point for client requests. API Gateway can serve as the reverse proxy to redirect all client requests to internal services. This endpoint is exposed to the world and can aggregate responses and send them back as one response, optimizing the number of calls.
- **Microservices**: API Gateway is a standard for managing microservice environments and is a go-to service for organizations transforming their monoliths to microservices.

- **Initiating API approaches**: Instead of building a full-blown API management platform, it is easy to get started with API Gateway and experience how easy it is to set up and use. It also offers a good business case for *build versus buy* to upgrade your legacy applications with a gateway implementation.

With the flexibility to simultaneously use the previous API versions even after the later versions are published, API Gateway can help manage multiple release versions. Backward compatibility is something that you should look into as your teams adopt services like this, and API Gateway not only offers that but also reduces cross-cutting concerns.

Amazon EventBridge

Amazon EventBridge is a fully managed event bus that can be used to design and build event-driven applications at scale. At the core, it is a serverless event bus that can be used to publish events and subscribe to events. The events invoked from your applications or SaaS products and even AWS services can be easily published in EventBridge. This service handles rule-driven routing effectively to various targets, enabling multi-service decoupled communications.

Features

EventBridge offers many features that make it best-in-class. Let us take a look at some of its important features:

- Built-in schema registry for automatic schema discovery and API destinations to minimize custom code

- Fully managed and scalable event bus without having to manage the infrastructure or capacity provisioning

- Global endpoints, enabling inter-service communication

- More than 100 built-in sources and targets, including AWS Lambda, Amazon SQS, Amazon SNS, and so on

- Robust event filtering and reliable delivery of events

- SaaS integration with more than 35 SaaS providers to manage seamless authentication while integrating

- Integration with the AWS ecosystem for monitoring, auditing, security, and compliance

EventBridge supports a number of SaaS platforms for integration, which opens the door for a variety of use cases, whether it is for enterprises to use them as building blocks for securing their applications, such as Auth0, or industry leaders such as Blitline to build massive document processing services.

Use cases

EventBridge is ideal for event-driven applications at scale. The following are some of the primary use cases of EventBridge:

- **Decoupling microservices**: EventBridge is a great option for implementing asynchronous messaging across your application in order to decouple independent components. EventBridge can act as the middle layer to receive event messages and handle routing to the respective services. Examples include an e-commerce backend system where a buyer makes a payment and the logistics department needs to be notified to keep the inventory in sync.

- **Serverless**: Events and serverless go together very well. Designing your serverless applications with Amazon EventBridge standardizes best practices for service integration, making it easy to consume or build a new service.

- **Connecting different services**: If you have applications that are on multiple different platforms and disjointed, EventBridge can be leveraged to keep them in sync. Changes in one system can be pushed to the event bus and processed by systems that depend on that information.

Amazon SNS

Amazon **Simple Notification Service (SNS)** is a managed service with which you can send messages from publishers to subscribers. The communication is asynchronous, and the consumers can subscribe to SNS, thereby enabling applications, end users, and devices to instantly receive notifications.

Features

Amazon SNS has multiple features and capabilities, including the following highlights:

- **Messaging**: Application-to-application messaging is supported by SNS with subscribers such as AWS Lambda functions, Amazon SQS queues, and HTTP/S endpoints.

- **Notifications**: Application-to-person notifications are supported where mobile applications, phone numbers, and email addresses can be verified subscribers.

- **Topics**: The **first in first out** (**FIFO**) topic is supported to ensure message ordering is not faltered.

- **Message durability, security, archival, filtering, and analytics** are all supported by SNS.

SNS is a pay-as-you-use service where a free tier is included when you first get started and you then incur charges upon reaching 1 million Amazon SNS requests.

Use cases

Application-to-application (**A2A**) and **application-to-person** (**A2P**) communication is fully managed by SNS through publisher/subscriber functionality. SNS is the best fit for distributed systems, microservices, and event-driven serverless applications where the publisher systems can fan out messages to a number of subscriber systems:

- **Application alerts**: If you have applications that need to trigger alerts based on predefined thresholds, then SNS can send these alerts to registered users via email or text messages.

- **User notifications**: If you have applications that need to send push-based email notifications or text messages, then SNS can be used.

- **Mobile push notifications**: If you have applications that need to send messages directly to mobile applications, then SNS can be used to send notifications to an app.

- **Fan-out to external targets**: Fan-out scenarios are applicable whenever messages need to be pushed to multiple endpoints such as HTTP(s) endpoints, Lambda functions, Amazon SQS queues, or Kinesis Data Firehose delivery streams. SNS supports parallel asynchronous processing.

- Secure delivery of messages, using message data protection that provides governance, compliance, and auditing services for enterprise applications that are message-centric to grant different content-access permissions to individual subscribers.

SNS is a HIPAA-eligible service so it can be used to build HIPAA-compliant applications if you execute a **Business Associate Agreement** (**BAA**) with AWS.

Amazon SQS

Simple Queue Service (**SQS**) is a message queuing service used to decouple and scale applications. A fully managed service, SQS is a highly scalable, reliable, and durable message queuing service that can be used to reduce the complexity and overhead associated with managing and operating message-oriented middleware.

Features

SQS can act as a messaging backbone and help decouple dependencies for microservices, distributed systems, and serverless applications. Here are some of its features:

- Supports SQS standard queues for maximum throughput, best-effort ordering, and at-least-once delivery

- Supports FIFO queues for messages to be processed in the exact order they were sent

- Seamless integration with AWS services such as Amazon Redshift, DynamoDB, RDS, EC2, ECS, Lambda, and S3 to scale distributed applications

- Supports batch operations to smooth out volume spikes without losing messages or increasing latency

With SQS, you get standard queues and FIFO queues. The standard queue provides maximum throughput and at-least-once delivery and best-effort ordering. FIFO queues guarantee message delivery exactly once in the exact order that they were sent. The benefit of using SQS is that AWS takes care of the ongoing operations and underlying infrastructure with no upfront cost or requirement to build out the supporting infrastructure.

Use cases

SQS helps address many real-time business scenarios where operational overhead can be significantly reduced, such as the following:

- **Decoupling distributed systems**: Cloud-native solutions where the frontend is separate from the backend can benefit from SQS to decouple incoming jobs from pipeline processes.

- **Handling message delivery without interruptions**: Many businesses experience losing messages between integrated systems during maintenance windows. With SQS, messaging infrastructure can be automated, fully reducing the need for maintenance operations. In addition, when messages are sent to SQS, there are parameters to adjust delay times so that messages that encounter errors during maintenance windows can be resubmitted.

- **Handling concurrency**: Many businesses encounter scenarios where concurrency and strict ordering of messages is mandatory. SQS can help support simultaneous messages being sent to the consumer application and FIFO queues guarantee a strict order of message delivery.

- **Secure delivery of messages**: Businesses that need to adhere to compliance protocols require host-to-host secure channels. SQS integrates with AWS **Key Management Service** (**KMS**) to encrypt data at rest.

SNS versus SQS

SQS and SNS play a crucial role in allowing distributed systems to communicate with each other and support asynchronous orchestration between microservices. However, these services provide two very different functions, and it is easy to get confused about when it is appropriate to use one over the other. The following table demonstrates the differences and runs through some suitable use cases:

	SQS	SNS
Entity Type	Queue	Pub-Sub
Message Consumption	Pull mechanism	Push mechanism
Persistence	Messages are persisted. A 1-minute to 14-day retention period is supported.	No persistence
Consumer Type	All consumers need to be identical because messages are processed in the exact same way.	Fan-out is supported. Messages can be processed in different ways so multiple types of consumers are supported.
Use Cases	• Suitable for simple queue message requirements • Best for decoupling applications and concurrent asynchronous messaging • Only one subscriber is needed	• Suitable for publishing batch messages • Messages that need to be processed with multiple endpoints • Messages that need multiple types of subscribers

Table 8.4 – SQS versus SNS use cases

In summary, when you have to publish messages and deliver them to systems that need to be notified of an event, go for SNS. And when you have a delivery point where a sub-system can pick the messages and process it later, go for SQS.

Case study on AWS integration services

The following use case is a hypothetical case study that highlights how the usage of AWS integration services can help focus on innovation and develop with agility.

Business case/challenge

A New York based company shares hotel reviews across the globe. The reviews typically include information on amenities, services, and overall quality. The company published more than 1 million high-quality digital images, but the volume brought in operational overhead for the company to handle. They had to offload the content to a central repository on Amazon S3. S3 was an economical solution and the company had to rearchitect its application in order to be able to recreate images that would fit mobile and tablet devices. Typically, the reprocessing would take about 800 hours to complete using in-house solutions, which did not scale, along with the fact that the software had numerous bugs.

Modernizing using AWS integration services

With the images already stored in S3, the company came up with a new technology stack using Amazon EC2 to process the images. The company needed a cloud environment that could handle large processing jobs and, at the same time, scale down for smaller daily jobs. With a customized Amazon Linux AMI on EC2, the company connects to S3 and Amazon SQS using a Python interface to AWS. With the integrated workflow, and additional software running on EC2 to process the photos, the company was able to process thousands of photos each night using Amazon EC2 Spot instances. The company used SQS to communicate the photos that were ready for processing and the status of the job.

The company saw a 95 percent improvement when it came to the processing of photos. In addition, cost savings in capital expenditures made AWS a perfect match for the company's batch processing without worrying about machine expenditures or maintenance.

Introduction to AWS ALM services

Application lifecycle management (**ALM**) is a set of mechanisms that involve people, tools, and processes that are used to manage the lifecycle of an application. ALM plays a crucial role in rapidly improving the flow of changes to production and providing visibility while promoting communication and collaboration throughout the application lifecycle.

When it comes to building serverless applications on the AWS cloud, a recommended approach is to use **Infrastructure as Code** (**IaC**) for provisioning and managing cloud resources, which we will dive deep into in this section. IaC removes the possibility of errors that arise due to the manual deployment of applications. Given that it is easy to store IaC in a version control system, any changes can be tracked and audited easily.

Diving deep into AWS IaC tools

Two predominant services when it comes to automating the provisioning of infrastructure on AWS are **CloudFormation** (**CFN**) and AWS **Cloud Development Kit** (**CDK**). Let us look at what each of these services entails.

AWS CFN

CFN is an IaC service that is used to create, provision, and manage cloud resources and support your applications throughout their lifecycle. Developers get to create these JSON or YAML templates and provision cloud resources in a repeated, predictable, and safe manner, allowing for rollbacks and versioning. This convenient provisioning mechanism supports a broad range of AWS and third-party resources that are typically used by enterprise applications, including legacy applications built on AWS such as container-based or serverless applications.

As you create the code for your infrastructure from the available templates, developers can upload the templates into an Amazon S3 bucket. CFN then provisions and configures the resources specified in the template. The following visual representation shows sample CFN code and how the template is used to create AWS resources such as Amazon API Gateway, AWS Lambda functions, Amazon DynamoDB tables, and IAM roles:

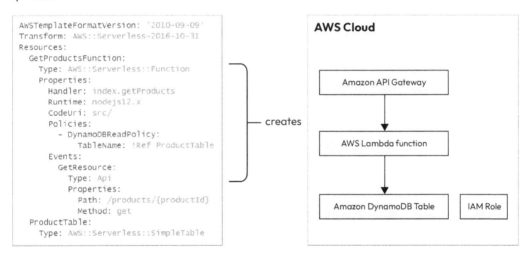

Figure 8.7 – CFN code example creating AWS infrastructure

One of the most popular functionalities for rapid infrastructure deployment has been CloudFormation stacks. Using CFN, developers can create an entire stack of infrastructure with one function call. In the following visual representation, a stack comprising multiple Amazon EC2 instances, EBS volumes, Elastic IP addresses, load balancers, Auto Scaling groups, RDS instances, SNS topics, Amazon CloudWatch alarms, Amazon SQS message queues, and Amazon SimpleDB domains is created.

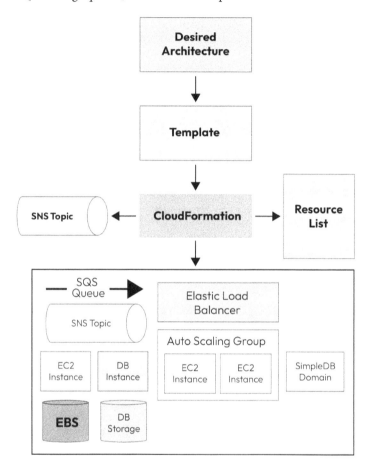

Figure 8.8 – CFN code example creating AWS infrastructure

There are four actions when it comes to the CFN stack: *Create*, *Update*, *Delete*, and *Roll back*.

CFN plays a crucial role in DevOps practices, simplifying provisioning and managing infrastructure, automating software release processes, and evolving your products at a faster pace than using traditional software development processes.

Features

CFN plays a powerful role in ALM and developers get to manage provisioned stacks that are fully automated instead of manually managing them. Let us take a look at some of the important features of CFN:

- **Authoring with JSON/YAML**: With the support of JSON and YAML, it is easy for users to model the entire infrastructure through templates. There is also a CFN designer IDE in the AWS console that you can use to style your infrastructure components visually.

- **Safety controls**: Changes are executed in a safe and controlled manner with automated provisioning where there are no manual steps to cause errors. In the event of an error, you can leverage rollback triggers to bring the whole stack operation to a previous deployable state.

- **Dependency management**: Developers can automatically manage dependencies through stack management operations such as create, update, delete, and roll back. CFN has the intelligence to determine the proper sequence of actions for every resource.

- **Cross-account and cross-region support**: You can provision AWS resources across multiple accounts and AWS regions with one CFN template.

- **Extensibility**: The ability to associate custom code provisioning and trigger Lambda functions through CFN is a powerful ability that CFN offers.

AWS CDK

As IaC becomes a more and more relevant and commonly used framework, developers need technologies and tools that are more developer-centric and easy to use. CDN does not offer built-in logic capabilities. CDK is one step closer to defining cloud infrastructure using programming languages such as C#, Java, Python, and Typescript.

CDK is an open source software development framework to define cloud application resources using programming languages. The familiarity and expressive power of programming languages makes CDK a favorable and safe platform to preconfigure cloud resources so you can build cloud applications with ease.

The following visual snippet is sample CDK code that creates a VPC on AWS:

```
export class NetworkStack extends cdk.Stack {
  public readonly vpc: ec2.Vpc;

  constructor(scope: cdk.Construct, id: string, props?: cdk.StackProps) {
    super(scope, id, props);

    this.vpc = new ec2.Vpc(this, 'VPC', {
      cidr: '10.0.0.0/16',
      natGatewaySubnets: {
        subnetName: 'Public'
      },
      subnetConfiguration: [
        {
          cidrMask: 26,
          name: 'Public',
          subnetType: ec2.SubnetType.PUBLIC
        },
        {
          name: 'Application',
          subnetType: ec2.SubnetType.PRIVATE
        },
        {
          cidrMask: 27,
          name: 'Database',
          subnetType: ec2.SubnetType.ISOLATED
        }
      ]
    });
```

Figure 8.9 – Sample CDK code to create a VPC on AWS

With AWS CDK, it is easy to share and reuse logic while defining your infrastructure. The commands that AWS CDK provides generate a CloudFormation template that you can integrate with the CI/CD process.

Benefits of CDK over CFN

CFN templates typically have low-level constructs that map one to one with CFN resources and attributes, while CDK introduces high-level constructs where you can build a set of resources with just a single line of code.

As a result, you get to write 50 lines of code in CDK to generate a CFN template of 1,000 lines of code. Let us look at the benefits of CDK over CFN in detail:

- **Easy to integrate with CI/CD processes**: With single commands, CDK can be used to define CI/CD pipelines using AWS DevOps tools such as Amazon CodeBuild and CodeDeploy.

- **Developer friendly**: As an extensible open source development framework, CDK offers **integrated development environment** (**IDE**) capabilities to build stacks. The developer's programming skills can be leveraged and any additional time it takes to learn new syntaxes can be avoided.

- **Easy to share and reuse**: Declarative infrastructure as code such as JSON or YAML is not something that developers are used to on a daily basis. The object-oriented language and the abstraction provided by CDK enable developers to define modules that can be shared across projects.

With AWS CDK, you can move several lines of code into abstractions, giving you the full power of the programming language to choose and define your application's infrastructure.

Case study on AWS IaC tools

The following case study is a hypothetical use case of web apps and microservices.

Business goal

A major insurance company needed to focus on three main areas for digital modernization: customers, agility, and cloud-native development. The company was incurring technical debt due to the compounded problem of using third-party tools and managing deployment and infrastructure. The company decided to revamp its existing deployment scripts as well as its applications.

Modernizing using AWS

In order to achieve the goal, the company made a strategic long-term business decision to go with a serverless-first approach. The motive is to gain an edge in a competitive, global, and increasingly digital market. Migrating the company's on-prem applications to the cloud and modernizing was a companywide goal.

Along with the serverless architecture, the company leveraged AWS CDN, to accelerate the definition of cloud application resources. The engineers found it easier to define these resources and did not have to write the code from scratch. With the help of templates, the teams were able to rapidly build projects and about 2,500 serverless patterns were deployed in one year by the company. As a result, the company was able to release high-quality solutions for customers on a faster timeline while reducing costs and removing infrastructure maintenance overhead.

Using AWS CDK, the company was able to migrate existing deployment and configuration scripts into a single code base and control flows using only a few lines of code. The company no longer had to create or maintain configuration files, which were often error-prone, and the overall code base size was smaller with improved readability.

Summary

In this chapter, we learned about key technical domains such as serverless, integration technologies, and the application lifecycle management framework to accelerate your modernization journey and implement an end-to-end seamless cloud architecture. We also learned about the different AWS services and technologies that can be leveraged for various use cases and would be a good fit for enterprises of any size.

In the next chapter, we will start a new technology discussion where we will learn how to modernize data on the cloud.

9

Modernizing Data and Analytics on AWS

The cloud has generated the next wave of invention that is driven by data. Organizations are striving to be data-driven and are looking for new approaches to managing exabyte-scale data and learning from it. Gartner's 2021 survey for Data indicates that organizations are increasingly deploying data and analytics solutions on the cloud to support digital transformation and acceleration. While they do so, questions such as "How do you transfer the data?" "How will the data be cataloged?" "How will you trace the data?" and "Who will be responsible for the data infrastructure platforms?" remain unfamiliar to organizations.

While the first step is to move the data to the cloud, it is not an easy feat, especially when it comes to keeping all the operations up and running while simultaneously moving data to the cloud. Choosing the right cloud partners and having a clear plan for migration and a vision is the right combination to execute modernization at every level. In this chapter, we will learn the why, what, and how of data modernization. Modern data architectures are not one-size-fits-all and often involve careful consideration of requirements such as ingesting data into a data lake, analyzing data from disparate data sources, complex interleaving and cleaning of data coming from time-based or disparate data sources, and so on.

We will cover the following main topics and explore a collection of AWS services so you can plan your modernization efforts and build state-of-the-art data platforms:

- Introduction to data infrastructure modernization
- Strategies for data modernization on AWS
- Modernizing databases on AWS
- Data case studies on AWS
- Introduction to analytics modernization
- Analytics case studies on AWS

Introducing data infrastructure modernization

For today's businesses, modern data architecture plays an important role in driving intelligence into their actions as part of the strategy for revenue growth. Digital transformation can be perceived as a trifecta: technology, business processes, and people working together. Getting the right data to the right applications at the right moment is always a challenge, especially with retro-fitted data management infrastructures that organizations may have inherited from several years of running legacy platforms. In this section, we will learn what data infrastructure modernization is and why IT leaders should prioritize it. We will also look into the key challenges involved in data modernization and the key steps in modernizing a legacy data system.

The following are a few of the challenges that organizations face while managing their enterprise data:

- More data than ever is being generated

- Data is being stored in silos across multiple data stores

- **Machine learning** (ML) adoption is challenged by a lack of skills and organizational inertia

- Data security, privacy, and compliance regulations are becoming increasingly important

To survive these challenges, organizations are making decisions to adapt to the rapidly evolving market dynamics with great urgency and their customer preferences. Data is critical to making faster and more informed decisions. As a result, enterprises are modernizing their data infrastructure and inventing new customer experiences.

> **Data infrastructure modernization**
>
> Building a solid foundation for your enterprise is key to unlocking attributes such as scale, performance, unified data access, security, governance, and AI/ML to solve today's and tomorrow's business challenges.

You may still be wondering about the appeal of data modernization and how it can empower more companies to build insights efficiently with the right data governance in place. Let us take a look at some of the benefits of data modernization.

Benefits of data modernization on the cloud

Data modernization allows businesses to develop systems and processes that can accelerate revenue. Cloud-based data platforms enable you to accelerate your data modernization journey through several advantages, such as elasticity, capacity, data pipeline automation, and advanced tooling. Here are some of the benefits that data modernization brings:

- **Increased data processing speed and efficiency**: Data is the core focus for organizations for any decisions that need to be made. Accurately analyzing data is a common challenge that enterprises face. Manual data entries, analysis, and improper data cleansing can hinder your data processing strategies. Data modernization dramatically reduces the total time it takes for data scientists to capture insights, patterns, and high-value data. Cloud platforms enable the automation of these processes by helping data scientists to seamlessly scale without having to worry about any manual or standard database operations.

- **Improved decision-making and deeper insights**: It is easier to make informed decisions when data scientists have real-time access to the data. With the self-service data platform model, it just becomes easier for the data teams to collaborate, innovate, and get deeper, quicker, and better insights.

- **Improved reliability**: By the end of 2022, there will be 97 zettabytes of data in the world (`https://webtribunal.net/blog/how-much-data-is-created-every-day/`). Legacy data architectures create an overhead of challenges during ingesting and processing. Data modernization enables consolidating data across your organization, which not only eliminates data silos but also facilitates a unified data view across different teams with cross-business functions.

- **Improved data governance and security**: The need for data governance and security is rising, especially for government and banking sectors where IT security and compliance are significant requirements. Companies must withstand cyber threats along with aligning to the changing laws and regulations. Data modernization can help improve data security with mechanisms such as encryption in transit and at rest.

Real-time network security and monitoring services available with the cloud vendor's ecosystem enable you to proactively detect and mitigate any attacks or threats. This also helps in improving data sovereignty requirements.

While getting off on-premises databases and moving to cloud data infrastructure is one step, there are different scenarios that you can incorporate to effectively implement data modernization. We will learn more about them in the next section.

Strategies for data modernization on AWS

AWS provides a reliable, secure, and scalable platform with the most comprehensive database services and solutions to transform your data infrastructure. As IT leaders, you get to empower your teams to build modern applications that can easily support the growing data on modern infrastructure. Let us learn about the strategies you can leverage for data modernization on AWS.

Break free from legacy databases

In the past few decades, enterprises of all sizes have been building and running their applications with dependencies on commercial relational databases. Purchasing long-term contracts and the database administrators getting skilled in the technology is a must to ensure no interruptions. With the times changing and the cloud opening up options for building and managing your applications and infrastructure, you no longer have to depend on legacy commercial databases. With the cloud benefits that we learned in the prior section, you don't have to be locked into long-term contracts with these commercial database vendors either.

Let us take a look at a few reasons why a cloud-native database will give you a better **return on investment** (**ROI**) than legacy databases:

- **Cutting costs on licenses and operating expenses**: Companies have to pay heavy bills to maintain the annual licenses of commercial relational databases. These license agreements are often stringent in terms of usage and impose several restrictions on your needs and costs. With cloud-native databases, there are no licenses, and you pay for your usage. This results in no capital expenditure and reasonable operational expenditure, with scaling as the biggest benefit.

- **Improved performance with cloud-native design**: With commercial databases, you need to perform capacity planning and provision your databases as per your analysis. Traffic instability results in many enterprises overpaying for unused resources or performance bottlenecks.

  Cloud-native databases are built to meet the dynamic requirements of today's digital business needs and can scale up to handle traffic bursts gracefully. In addition to the enormous scale, databases spanning multiple availability zones ensure maximum availability and durability.

- **Foster agile development**: New requirements and changing market conditions drive continuous iterations. We learned in *Chapter 5, Modernization on the Cloud*, how microservices architecture facilitates developer agility in such scenarios. The modular nature of microservices results in applications being broken down into smaller pieces that can be deployed at scale. Such microservices architectures require asynchronous communication with databases upon data changes in your system. Modern databases make it easy to implement microservices with data richness as the core focus.

- **Reliability**: This is another key differentiator that makes fully managed databases on the cloud ideal for running business-critical workloads in a highly reliable manner. Fully managed databases have features such as multi-region capabilities, network isolation, visibility, and end-to-end encryption baked into their offering. This makes it highly compelling for businesses to leverage cloud databases for their workloads. The support for continuous monitoring and automated scaling also helps you with self-healing.

As an IT leader, you are the key force behind data modernization; you need to have a detailed understanding of the options that are available to move from a legacy database to a cloud-native database. This will enable you to break free from legacy databases and become equipped to stay aligned with the latest technologies.

When migrating from legacy databases to cloud-native, it is important to research your options and carefully understand the following factors:

- **Choosing the database type**: Picking the right database type for your application is the first and foremost aspect. The dominance of relational databases is gradually reducing with the arrival of NoSQL databases. You are no longer constrained by strict schema requirements, foreign key/primary key validations, or SQL operations if your applications don't require it. The fundamental data modeling principles vary for each database type, so understanding the benefits and the nature of your application design and how the data flows through will help you rearchitect your data model. Skill set is also a key factor to remember with database architects focusing on specific technologies as the domain expands.

- **Choosing the migration process**: Most cloud vendors offer self-service options for migrating from a self-managed database to a fully managed one. These services let you migrate data between homogeneous and heterogeneous database types, making it easy for you to convert from relational to non-relational. If you have complex data models and do not have the right skill set to decide upon these factors, cloud vendors also provide expert advice through professional service teams who can help you plan and successfully execute your migration. For example, AWS offers to securely discover, assess, convert, and migrate database workloads with automated migration.

Building modern applications with purpose-built databases

Data models have been perceived as hierarchical, where the parent and child records are connected with limited abilities to accommodate complex concepts, such as many-to-many relationships. With the introduction of web-enabled global businesses, we have seen a dramatic shift in the database landscape.

Developers are building faster, more scalable, and more resilient applications to meet the changes and demands. Relational databases don't always fit perfectly to meet this rapid shift and are pushed to their limits, especially schema enforcement and transaction support. The need for storing heterogeneous data records in various data stores, such as document-based or graph-based, is growing. Provisioning data stores for each of your business applications is a non-trivial task, and many cloud providers have built purpose-built databases to cater to this requirement and seamlessly enable modern application development.

Moving to managed databases

The *lift and shift* strategy is the easiest first step toward the cloud. Although you may see benefits from the cloud by doing a simple lift and shift of your databases, you are still bound to have operational costs. Moving to higher-level services such as serverless or fully managed databases can enable you to unlock the key benefits of modernization, such as lower IT spending, innovation, and performance, all at zero operational overhead.

With the rapid shift of application workloads from on-premises locations to cloud-based computing options, companies must look for simple data and storage options. A recent study by the **International Data Corporation** (**IDC**) (`https://pages.awscloud.com/GC-600-business-value-rds-ebook-learn.html`) indicates that you can achieve 86 percent faster deployment of new databases, 97 percent less unplanned downtime, and a 264 percent ROI over 3 years by using high-level services such as fully managed databases. Let us learn about the key differentiating factors:

- **Database management**: Database management is considered heavy lifting involving fundamental operations such as managing backups, building tooling for monitoring, patching, minor upgrades, and having 24x7 support. This involves hiring database administrators who are experienced and possess the right skill set.

- **Failover and recovery**: Planning for disaster recovery is a non-trivial task where the operating staff must have tried and tested runbooks ready. Operating on these runbooks can often be stressful for these operators. Downtime is expensive to businesses, and a faulty backup plan can result in catastrophic data loss.

- **Flexible compute for capacity planning and scaling**: Capacity planning is an aspect that needs regular touch points if you are managing on-prem data centers. Guessing the growth rate of your business and planning for the required compute accordingly often results in overestimations. Wasting your money on unused infrastructure leaves businesses in complex negotiation scenarios with their vendors.

In the next section, you will learn about different database service engines available on AWS and how you can leverage these cloud-native databases.

Modernizing data using AWS

Modernizing your data warehouses and infrastructure can be a multi-year journey involving many teams. There are several factors involved in making an informed choice of the data model you need for your applications and the database technology to rely on. This is an early decision, keeping in mind the long-term impact you will drive from the decision. In this section, you will learn about different data models, how to choose the right data model for your application, and AWS's data ecosystem.

Choosing the right database type

The right tool for the right job is a key strategy to be successful in data transformation efforts. There are many options to choose from when it comes to cloud-native databases. Choosing the right database is a decision that needs to consider several factors such as *access characteristics, application performance, type of operations, and data patterns*. The following are some guidelines you can consider when picking the right type of database.

Workload type

Application type defines the nature of the workload and the data access patterns. Regarding the type of workload, there are three types: *transactional, analytical, and cache-based*. When you understand what type of workloads you are running, it will help you choose the database type:

- **Transactional**: This refers to applications requiring a high number of concurrent operations and where each operation needs to be sequential and is reading or writing data. Also called **online transactional processing** (**OLTP**), this applies to web applications for e-commerce, gaming, and social networking.

- **Analytical**: This refers to applications that are characterized by aggregates of large volumes of data. Also called **online analytical processing** (**OLAP**), access patterns are for analytical data stores and used for reporting.

- **Cache-based**: This refers to applications where frequently accessed data needs to be stored in a different datastore for low latencies. The cache is usually used for faster response times to reduce the load on transactional databases serving as a secondary source of data in conjunction with transactional workloads.

Data structure

The data format structure is important to know to help define the access patterns. When it comes to the data shape, there are a few common data models in which the data is structured, such as SQL schema, semi-structured, such as JSON objects, or no-structure, such as textual data, or a key-value pair, such as a filename to file content relationship.

So, we broadly classify the data models into *relational, key-value, document, in-memory, graph, time-series, and ledger*. Let us learn some details about each model.

A **relational** database is the most common database model, where data is normalized into tables and related data is stored together to be easily queried via **structured query language** (**SQL**). The schema validation on this data model is strictly enforced to ensure data integrity for your applications.

The following diagram is a model of how relational database tables are structured:

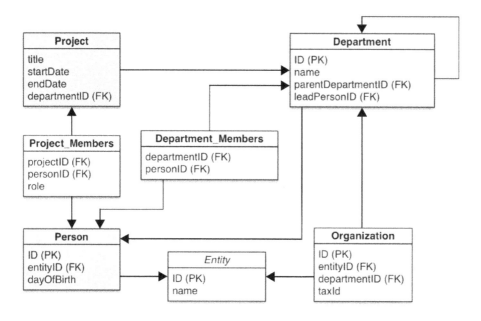

Figure 9.1 – A relational data model

AWS offers services such as Amazon **Relational Database Service** (**RDS**) and Amazon **Aurora** for applications to rely on and run high-performance applications for relational data models. RDS and Aurora are fully managed services.

The **key-value and document** model, also called the wide-column data model, is for applications that need high scaling capabilities. As your data grows, it is often stored as key-value data. The data is broken into partitionable chunks and stored to accommodate infinite scaling at high performance that other data models may not help you achieve. Gaming, **Internet of things** (**IoT**) devices, and tech-based applications can benefit from these data models.

The **document** database model is used to assemble frequently accessed heterogeneous data records also called documents. This is especially useful for providing faster queries, handling big data, and a simplified method for maintaining databases. The data is kept together in the form of a document for faster access. Data is represented as a JSON document, as shown in *Figure 9.2*:

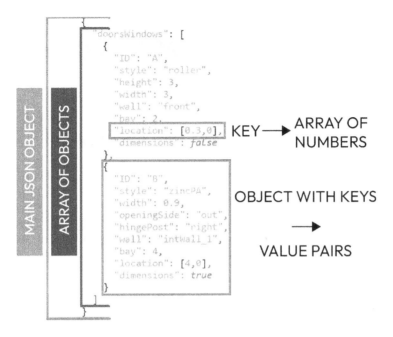

```
"doorsWindows": [
  {
    "ID": "A",
    "style": "roller",
    "height": 3,
    "width": 3,
    "wall": "front",
    "bay": 2,
    "location": [0.3,0],     KEY ——→ ARRAY OF
    "dimensions": false                 NUMBERS
  },
  {
    "ID": "B",
    "style": "zincPA",
    "width": 0.9,                OBJECT WITH KEYS
    "openingSide": "out",
    "hingePost": "right",            ——→
    "wall": "intWall_1",
    "bay": 4,                     VALUE PAIRS
    "location": [4,0],
    "dimensions": true
  }
]
```

Figure 9.2 – The key-value and document database model

Amazon DynamoDB is a document database that accommodates key-value data storage and achieves millisecond performance at any scale.

The **graph** data model highlights the data relationships to capture the hidden connections between data. This database type stores nodes and relationships instead of tables or documents. Relational databases can store relationships, but you must navigate the connections with JOIN operations that are usually expensive on cross-lookups.

A graph database enables faster traversing through data at a million connections per second rate, which makes it desirable to solve modern data problems such as navigating through data hierarchies, finding hidden connections between unrelated or distant items, and discovering relationships between items. The following diagram is an example of how social network connections are defined. Each node represents a person, and the edges are the relationships between the people. Through this data model, it is easy to discover the *friend of friends* for a given person:

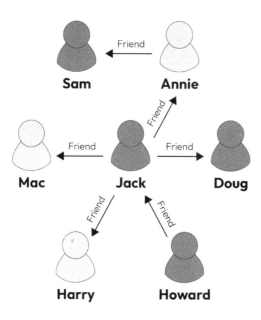

Figure 9.3 – A graph data model

Amazon Neptune is a graph database engine that is fully managed and is ideal for storing relationships and querying the graph within milliseconds. Neptune facilitates the storage of billions of records and supports many popular graph models defined by the **World Wide Web Consortium** (**W3C**) **resource description framework** (**RDF**), making it easy for enterprises to define and configure industry standard patterns.

In-memory databases work best for applications with use cases such as leaderboards, session stores, and real-time analytics. The in-memory databases store the data in memory for faster operations. With microsecond response times, these databases facilitate rapid disk access compared to **solid-state drives** (**SSD**) or **random access memory** (**RAM**). These databases have a high risk of losing data in the case of a server failure, so typically, AWS takes snapshots for each data storing operation along with storing it in a log.

In the following diagram, the primary node uses an in-memory storage engine, and it replicates two other secondary nodes. In the case of a failure, the secondary would become the primary in-memory server, thereby retaining quick access to the data:

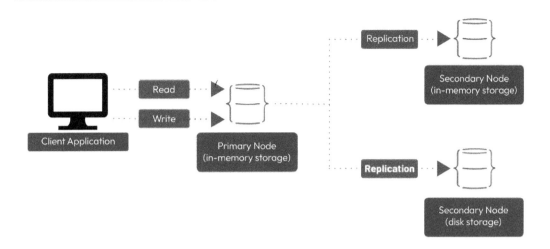

Figure 9.4 – An in-memory data model

Some examples of in-memory databases on AWS are *Amazon MemoryDB for Redis, Amazon ElastiCache for Redis, and Amazon ElastiCache for Memcached*. These databases enable ultra-fast performance to power internet-scale and real-time applications.

Search engine databases use indexes on data samples to capture and categorize characteristics that are similar to searching for data content. Data can be long, semi-structured, or unstructured to help cost-effectively retrieve high-quality information. Highly optimized for keyword queries, these are ideal for full-text searches, complex search expressions, and ranking of search result use cases. The following sample shows how data is organized by relevance, which can include any type of data, such as text documents, geographic information, images, video or audio content, and any other data payload:

Index	Schema			
Id	Title	Release_year	Rating	Genre
1002	Star Wars: Episode V The Empire Strikes Back	1980	8.7	Action
1003	The Godfather	1972	9.2	Drama
1004	Mission Impossible: 2	2000	6.1	Action
1005	Alive	2020	6.2	Action

Figure 9.5 – A search-engine data store example

The **Amazon Elasticsearch** service provides APIs and real-time search, monitoring, and analytics for every industry such as health care, telecommunications, retail and e-commerce, technology, and financial services. It provides search-powered capabilities such as capabilities to perform activities such as tracking down content with internet searches, improves observability, and builds better applications with interactive log analytics and real-time application monitoring.

The **time-series** database service stores and retrieves data records that are part of a time series and have associated timestamps. The timestamp provides critical context and is useful for complex analysis of the data. This database model allows you to store large volumes of timestamped data in an easy format for inserts and retrievals. The following example shows a time-series database containing data for each point in time:

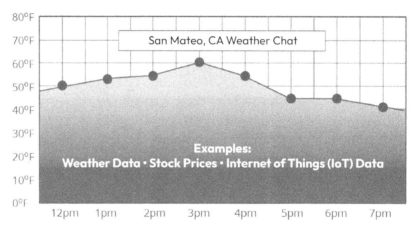

A Time Series Database is a database that
contains data for each point in time.

Figure 9.6 – Time series data model

Amazon Timestream is a time-series database that can store and analyze trillions of timestamped events. This is a serverless database, and you don't have to manage the underlying infrastructure or adjust the compute for capacity and performance.

The **ledger-based** database provides an immutable data store and a transaction log of any changes. This is especially useful in system failures, disaster recovery, or even data replication when you need to replay transactions. However, the transaction logs are not immutable and will not provide direct access to end users. Amazon **Quantum Ledger Database** (**QLDB**) is a fully managed database to track all application data changes and maintain a transparent and cryptographically verified transaction log.

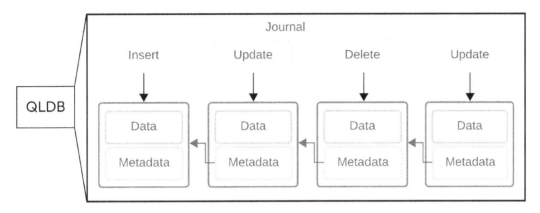

Figure 9.7 – A QLDB data model

As shown in the preceding diagram, an application is connected to a ledger and runs **create, read, update, and delete (CRUD)** operations, which make changes to a table named cars. You will notice that the data is written to the journal in a sequence. The data is then formulated into the table with views of the current state and complete history, along with a version number.

While we looked at a variety of databases: relational, document, key-value, graph, in-memory, and search, the following rubric will help you get a single view of all the AWS offerings that are available so you can find the right choice:

Database Type	AWS Service	Characteristics	Use Cases
Relational	Amazon Aurora Amazon RDS Amazon Redshift	**Atomicity, consistency, isolation, and durability** (**ACID**) transactions, schema on-write, referential integrity, and secure and stable	Enterprise-level applications, and online transaction systems
Key-value	Amazon DynamoDB	High throughput, low latency reads and writes, and high scaling capabilities	Shopping cart data and leaderboards
Document	Amazon DocumentDB MongoDB on AWS	JSON-styled, document-oriented, and a flexible schema	Customer data, user-generated content, and order data
Wide column	Amazon Keyspaces	Schema agnostic and handles querying non-sequential data	Web analytics and analyzing data from sensors

Database Type	AWS Service	Characteristics	Use Cases
Graph	Amazon Neptune	Stores relationship between diverse data points in the form of nodes. Most scalable	Data visualization purposes and fraud prevention
Time series	Amazon Timestream	Built-in time-series analytics for near real-time	IoT, analytics, and DevOps applications
Ledger	Amazon Ledger Database service (QLDB)	Transparent, immutable, and cryptographically verifiable transaction log	Storing financial transactions, reconciling supply chain systems, maintaining claims history, and centralizing digital records
In-memory	Amazon ElastiCache Amazon MemoryDB for Redis	Submillisecond latency and key-value pair cache	Gaming leaderboards, messaging, and pattern matching

Table 9.1 – Database model types and use cases

As you look at this landscape of database options, it is important to capture all the requirements when making a decision. If you have a specific use case or a business challenge in mind, pick two or three options and drill down by each of the specific problems that you want to solve, and use the preceding rubric to determine the best choice.

Performance needs

Understanding the *performance requirements* of your application in terms of how fast your data access needs to happen based on the size of the data records is an important aspect. When running mission-critical workloads such as voice, video, and even the evolving metaverse, *zero latency* is of great importance. Using an in-memory cache in those scenarios is ideal.

If you are running *analytical workloads* where background job processing and millisecond latencies are acceptable, you should focus on how you build the service to handle the intake of the amount of data into your system.

The *geographic placement* of the data is crucial for lower response times and compliance requirements. Therefore, considering the requirements for the replication of the data and how to handle it is key for such scenarios.

Operational overhead

The operational burden of the databases is half of the battle when it comes to your teams ensuring failovers, configuring backups, and plans for upgrades, as we discussed previously. Moving from self-managed databases to purpose-built databases enables you to lift the operational burden from you. For AWS, all the purpose-built databases take care of this aspect for you.

Data case studies on AWS

In this section, I will introduce you to a few organizations that have successfully transformed their business processes. These businesses were able to enhance their customer's experience by modernizing their legacy data platforms and transforming to leverage fully managed database AWS services that we learned about in the previous sections.

Case study – Amazon Aurora

This case study focuses on how businesses can evolve their business applications using Amazon Aurora.

The business case and challenge

A leading online travel web portal, *TRVZ*, operates across 70 countries and decided to innovate its global payments system. TRVZ caters to both consumers' and partners' payments across their global platform business. As the company witnessed an increase in traffic, transactions took longer to run and created a cascading effect of performance and latency issues. TRVZ acquired a commercial database, SQL server, and over the years, it was getting difficult to scale up every time the traffic requirements changed.

To build a portal to meet high traffic and availability requirements, TRVZ migrated to a microservices-driven architecture using AWS infrastructure. Working with a single, large relational database and having one large schema impacted TRVZ negatively with consistency and reliability issues.

Modernizing on AWS using Amazon Aurora and Amazon Elasticsearch Service

TRVZ decided to modernize its legacy database system and move it to Amazon Aurora, a MySQL and PostgreSQL compatible relational database, given its cost and performance benefits. The database developers refactored their existing SQL Server schema to Aurora PostgreSQL using the **AWS Schema Conversion Tool** (**AWS SCT**). This tool makes the conversion of the existing database schema and database code objects, including views and stored procedures, from one database to another easier.

Database migration, including the schema update, took about a week to develop and test. The development team started migrating their microservice components built on a Java framework using Spring Boot and now has about 15 microservices built on the Spring stack.

Additionally, TRVZ enhanced its travel portal to provide real-time data to its users and operations teams by using Amazon Elasticsearch Service. This enabled TRVZ's users to search, analyze, and visualize travel booking costs effectively. With the microservices-driven architecture, TRVZ was able to scale up and handle sudden traffic bursts and scale down when traffic subsided. With the fully managed database, TRVZ could sustain 500 transactions per second with zero time on maintenance and no in-house database administrators.

Case study – Amazon DynamoDB

This case study is to provide information on how businesses can transform their applications using Amazon DynamoDB.

The business case and challenge

McSec is a leading global company helping consumers and businesses with antivirus software, **virtual private networks** (**VPNs**), and identity monitoring. As the applications became vital for their business, McSec aimed to meet resiliency requirements and manage the capacity demands to achieve high performance and improve operational efficiency. The company was operating its legacy application using on-prem infrastructure. The commercial database that McSec was using did not scale well enough, and the company realized that it had to refresh its relational database management system.

Modernizing on AWS using Amazon DynamoDB

The company started its modernization efforts by evaluating the AWS database service offerings keeping in mind scalability and high-performance requirements along with its interest in choosing a key-value NoSQL database. It opted for Amazon DynamoDB with provisioned throughput capacity to enable the other applications that use the same database to horizontally scale and adjust the throughput capacity in response to actual traffic patterns. This helped them bring down response times and handle 5 billion read operations per month and 3 billion write operations with 40 percent savings in their monthly infrastructure costs.

To improve their telemetry, McSec used Amazon CloudWatch to help increase monitoring and observability for its site reliability engineers. This gave them additional metrics to monitor response times and helped them stay focused on business objectives. This encouraged McSec to expand its infrastructure on the cloud and enable more operational excellence overall.

Case study – Amazon DocumentDB

This case study is to showcase how businesses can transform their applications using Amazon DocumentDB.

The business case and challenge

A unicorn start-up company, an app-based marketplace to deliver everything from groceries to clothes, recently earned an additional investment to expand its services into four other countries. The company initially built a monolithic application with traditional architectural design patterns. With the increased expansion, it determined that it needed a database platform that could support microservices-architecture and facilitate its innovation to compete in the on-demand delivery market.

Modernizing on AWS using Amazon DocumentDB

The company required multiple databases for different functionalities. It needed to facilitate functionalities such as catalog search, improved performance, and reduced latencies. They decided to use Elasticsearch and set up a series of microservices organized by country where each country is supported by one Elasticsearch cluster. They wanted to keep the price tag low by spinning up several nodes per cluster. The company moved from Elasticsearch into document store solutions and needed a fully managed document database service.

After migrating to Amazon DocumentDB, it was able to run thousands of different databases in production, and the new model allotted one DocumentDB cluster per country with much fewer configuration changes. The company achieved better performance, less maintenance, and firefighting, and was easily able to handle the surge in their orders and deliveries, meeting the scalability and availability requirements.

Case study – Amazon Neptune

This case study is to provide information on how businesses can transform their applications using Amazon Neptune.

The business case and challenge

AnyCompany, which is a worldwide energy company, oversees a fleet of thousands of gas turbines. The company ran into challenges when trying to provision, scale, and manage clusters of database instances. It needed to manage a broad fleet of equipment across different environments and contexts, given that the business was growing rapidly. The company wanted to build applications and connected data points to query billions of relationships within the data for contact tracing of the fleets. It also needed a reliable way to model contextual data and complex hierarchies and enhance decision-making around its fleet of gas turbines. It quickly realized that doing this and managing such a broad fleet at scale was not a trivial job. After careful assessment, AnyCompany explored the knowledge graph technology as the ideal solution for organizing, managing, and querying its machine data structures across the fleet. Traditionally, businesses with unpredictable workloads must monitor and reconfigure capacity constantly to meet performance requirements.

AnyCompany built a core part of the turbine knowledge graph platform using Amazon Neptune and took advantage of its features to build visualizations of the complex relationships and connections within the fleet in question. The company was looking for ready-made elements to combine and configure for the data management tasks such as graph data visualization, search, editing, and so on.

Modeling the modern workplace by using a purpose-built graph database

AnyCompany decided to go with Amazon Neptune's graph database service with its unique advantages, such as durability, performance, availability, and ease of use. It decided to build its smart tool platform to efficiently manage the data. Deploying the knowledge graph application involved low-code development where huge amounts of structured and unstructured data from disparate data sources came in. AnyCompany also needed to generate relationship visualizations more intuitively. Using this platform, AnyCompany successfully deployed an end-user-oriented custom knowledge graph application for managing its fleet of large gas turbines in just a few months. From an operations standpoint, the company was able to turn around its development, testing, and deployment cycle by saving 1,000 hours of manual processing in the first year. Performance and agility helped them to go to market quickly without trading off any security standards or efficiencies.

Case study – Amazon QLDB

In this case study, we will explore how companies can reinvent their data platforms by using Amazon QLDB.

The business case and challenge

DeliveryIT is a consumer packaged goods delivery operations organization firm that provides quick, efficient, and cost-effective delivery solutions through its mobile application. The company faced several challenges in terms of lack of traceability, security, low transparency, and a manual acknowledgment system. The administration team faced a lot of delivery rejections, and the data collection system was unreliable. The company's client demands for secure delivery of confidential items were not being met. The company needed a safe way to deliver high-value objects and sealed documents with utmost security and verification.

Modeling a blockchain-based solution using Amazon QLDB

The company decided to build a tamper-proof secure blockchain-based QR code tracking system dependent on ledger-based technologies. Data integrity, completeness, and verifiability are critical to this platform and hence the company decided to go with Amazon QLBD. Getting a system of record applications and cryptographically verifiable history was core to implementing a centralized and transparent platform for DeliveryIT.

Each transaction is registered through a digital handshake, marking where and when exactly the items changed hands. DeliveryIT implemented the solution using this platform to meet mobile compatibility requirements so that when the items arrive at their destination, the mobile application can automatically generate a QR code. These QR codes are embedded with a hash containing metadata regarding the shipment, such as collection point, drop-off location, and delivery time. The customers acknowledge upon receiving the parcels by scanning the QR code, and the mobile application takes a photograph of the delivered package. The entire process serves as capturing the trail of the transaction from start to finish.

This helped improve security, transparency, and easy verification and tracking of shipments. The data was secure to be shared with all stakeholders across the logistics network to decrease the risk of conflict information.

Introducing analytics modernization

Data is the strongest asset for any business, and having a modern analytics ecosystem that is scalable, agile, and future-ready can help you stay ahead of how you can serve your customers. Analytics modernization is not just about moving to the cloud but rethinking how you use analytics as an organization. A balanced approach, including crucial aspects such as data strategy, data architecture, data management and governance, analytics tools, and the right people will help you to acknowledge a modern data architecture.

Data movement

Data explosion is prevalent in all organizations where the data volumes are growing dramatically from terabytes to petabytes and exabytes. When it comes to the data movement, here are the different types:

- **Inside out**: This is where you store data on-premise and move a portion of data to the purpose-built data stores for additional analytics

- **Outside in**: This is where you store data in purpose-built data stores and move the data to a data repository to run analysis on the data

- **Around the perimeter**: This is where you integrate data from on-premise and purpose-built data stores

- **Sharing across**: This is where you have modern data architecture to enable governance and data sharing across physical or logical boundaries

- **Data gravity**: This is where data is growing disproportionately across the purpose-built stores and on-premise, and it is important to use the right controls for analysis and insights

What does modern data architecture help with?

Organizations that want to go down the path of data modernization want to tackle several challenges when it comes to data and analytics. Here are the primary ones.

Data volumes

The exponential data volumes come with their own challenges, making it difficult for organizations to handle them effectively. There are competitive advantages that come with modern data architectures, such as the ability to easily access, use, store, transform, and analyze data.

Data types

With the microservices evolution, it is not just a relational data structure anymore. Different data models are evolving, as we discussed earlier, and there is always a need to bring these semi-structured and unstructured data sources together. The modern architectures for these data models will help bring integration and make it effective for businesses to adopt new mechanisms and technologies to harness all of the data easily.

Consumer types

There are different types of requirements and end-users for the data, such as reporting, dashboarding, streaming, embedded use cases, and advanced analytics. Modern data architectures enable data to flow multi-directionally throughout the life cycle.

Platform types

Many organizations are in the process of moving their systems to the cloud, and as a result, their data sources are split between on-prem, cloud, multiple clouds, or hybrid. This adds to the integration challenges. Modern data architectures help with providing integration services and management capabilities.

Advanced analytics

The need for advanced analytics is becoming more and more prevalent and having an ecosystem in place for ML, AI, and advanced analytics is desirable. Providing data access to more people means that there is also a need to provide the ability to slice and dice the data as per their needs. Modern data architecture lets organizations meet the rapidly changing demands in the marketplace.

Pillars of modern data architecture on AWS

The modern analytics management principles that the cloud incorporates often help you realize the potential beyond just cloud adoption, including abilities such as the following.

Scalable data lakes

A **data lake** is a centralized repository for you to store the data as-is at any scale. The data can be structured or unstructured and doesn't have to be preprocessed to fit into any format protocols. Building data lakes from scratch involves many tasks, such as data source identification, data ingestion, data preparation, preprocessing the data and cleaning it up, staging the data for queries, and incorporating ML for data lake analytics. This could take days to months for any organization.

AWS offers services such as Amazon S3 to build the data lake at scale with highly secure, durable, and available advanced security and audit capabilities. The following diagram illustrates the various data movement types we learned about in the previous section and how the data lake approach helps with the data in the real world between the data movement types:

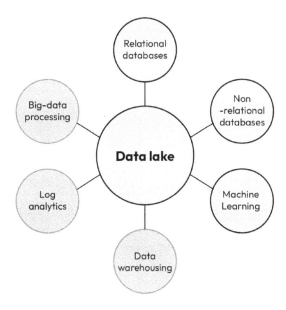

Figure 9.8 – Data lake approach

Implementing a data lake empowers you to learn, identify, and act upon opportunities that can accelerate your business growth. Here are some additional benefits:

- Improved business value
- Improved innovation
- Improved operational efficiencies

With the AWS offerings, you can build a cost-effective data lake in the cloud, analyze the data and incorporate a variety of analytical approaches, including ML. The following table describes five logical layers of a typical data lake and the specific AWS services that can help you execute the responsibilities listed in the following table.

The data lake architecture helps improve data ingestion from disparate data sources such as **Software-as-a-Service (SaaS)** applications, sensors, social media, web, and mobile applications.

Let us take a look at each of the layers and the AWS services that can help with data processes:

Layer name	Responsibilities	AWS service
Data consumption	Handles interactive SQL queries, warehouse-style analytics, **business intelligence (BI)** dashboards, and ML	Amazon Athena AWS Redshift with Spectrum Amazon QuickSight Amazon SageMaker
Processing	Provides multi-purpose-built components to enable use cases such as SQL-based **extract, transform, and load (ETL)**, big data processing, and near-real-time data streaming	Amazon EMR AWS Glue Amazon Kinesis Data Analytics
Catalog	Stores business and technical metadata about datasets. Enables writing queries and tracking versioned schemas and granular partitioning	Amazon Lake Formation
Storage	Provides durable, scalable, and cost-effective components to store and manage data	Amazon Redshift Amazon S3
Ingestion	Ingests data into the storage layer. Provides the ability to connect to internal and external data sources and deliver real-time streaming data into data warehouses, data lake components, and so on	AWS AppFlow AWS Data Migration Service Amazon DataSync Amazon Kinesis

Table 9.2 – Data layers and AWS services

A data lake implemented on AWS using the services listed in the preceding table helps you accelerate the modernization of your analytics along with the rich capabilities you can get with them. Choosing the right AWS service for the right job will be your responsibility to maximize the value from the available capabilities.

Purpose-built analytics services

The purpose-built analytics services are optimized to meet use cases such as analyzing data stored in data lakes, big data processing, interactive log analytics, and real-time video analytics.

Let us explore the use cases and features of each of the AWS services for analytics:

AWS service	Use cases	Features
Amazon Athena	<ul><li>Time-sensitive data queries</li><li>Archival log analysis and RTL</li><li>Data science and ML</li><li>ETL and reporting</li></ul>	<ul><li>Serverless, no need to manage infrastructure</li><li>Analyze unstructured, semi-structured, and structured data</li><li>Integrates with QuickSight for data visualization</li></ul>
Amazon EMR	<ul><li>Analyze and process big data or Hadoop-based processing</li><li>ML</li><li>Data transformations for ETL</li><li>Clickstream analysis</li><li>Real-time streaming</li><li>Interactive analysis</li></ul>	<ul><li>Leverage multiple data stores such as Amazon S3, **Hadoop distributed file system** (**HDFS**), and Amazon DynamoDB</li><li>Run deep learning and ML tools such as TensorFlow and Apache MXNet</li><li>Data pipeline development and data processing by using Apache Hudi, and Apache Spark through EMR Studio</li></ul>
Amazon OpenSearch Service	<ul><li>Log analytics and search</li><li>Real-time application monitoring and alerting</li><li>Ingestion of high volumes of log data for tracing</li><li>Asynchronous searches that can run in the background</li></ul>	<ul><li>Open distro for ElasticSearch SQL Engine</li><li>Interactive SQL **command-line interface** (**CLI**) and SQL Workbench that supports viewing and saving results in text, JSON, **Java database connectivity** (**JDBC**), or CSV</li><li>Connectors support for **Open Database Connectivity** (**ODBC**) and JDBC drivers</li></ul>

AWS service	Use cases	Features
Amazon Kinesis	<ul><li>Track real-time metrics</li><li>Run real-time audits</li><li>Create **Directed Acyclic Graphs** (**DAG**) of data streams</li><li>Real-time metrics extraction and report generation</li><li>Ingest and analyze real-time data, including video, audio, application logs, website clickstreams, and IoT telemetry</li></ul>	<ul><li>Video streams to allow secure streaming from video devices for analytics and ML</li><li>Data streams to allow real-time data streaming</li><li>Data Firehose to capture, transform, and load data streams for BI</li><li>Data analytics to process real-time data with SQL or Apache Flink</li></ul>
Amazon Redshift	<ul><li>Traditional data warehousing</li><li>Store and process log analysis, including spatial data</li><li>Analyze data for business applications</li></ul>	<ul><li>Federated query capabilities</li><li>Partner solutions integrations, such as Salesforce, Google Analytics, and Facebook Ads</li><li>ML and result caching</li></ul>

Table 9.3 – Analytics and use cases

These AWS services are all designed to maximize throughput and performance along with scale and cost efficiencies that come with the cloud benefits. All these features enable you to run petabyte-scale analysis at less than half the cost of running on-premise solutions.

Unified data access

It is important that you can move data from one store to a unified data store or data lake. Here are a couple of AWS services that provide this function.

AWS Glue

AWS Glue is a serverless ETL service that can be used to discover, prepare, and process data for analytics, ML, and building your applications. Glue provides data integration capabilities that are often instrumental to executing tasks performed in data warehouses and data lakes. Tasks such as data discovery, data extraction, and enriching, cleaning, and normalizing constitute data integration processes. There are several AWS services that AWS Glue supports, such as Amazon Aurora, Amazon RDS for MySQL, Amazon RDS for Oracle, Amazon RDS for PostgreSQL, Amazon RDS for SQL Server, Amazon Redshift, DynamoDB, and Amazon S3, as well as MySQL, Oracle, Microsoft SQL Server, and PostgreSQL databases in your VPC running on Amazon EC2. Glue also supports data streams from Amazon MSK, Amazon Kinesis Data Streams, and Apache Kafka.

Components of AWS Glue

The AWS Glue ecosystem has the following components to achieve the work that is required to ETL data from a data source:

- The data catalog holds the metadata about the data such as data sources, destinations, required transformations, queries, schemas, partitions, and so on.

- Data crawlers and classifiers enable you to scan repositories, classify them, and extract the information. This data is then stored in the data catalog to guide the ETL operations.

- The console lets you define jobs, connections, tables, and scheduler setup and provides access to API endpoints that are used to programmatically interface with the backend.

- The job scheduling system orchestrates your ETL workflow and automates the scripts to extract, transform, and transfer data to other locations.

- Streaming support provides sources such as Apache Kafka or Amazon Kinesis to ingest, process, and enrich the data on the go. This enables you to work simultaneously with multiple data sources, such as streaming and batch.

Glue is a serverless ETL service that enables you to be cost-effective and take advantage of using it in your own organization. Let us take a look at some of the real-world examples in the next section.

Use cases

Glue is an effective tool for ETL task orchestration. It integrates well with AWS services such as Amazon Kinesis, Amazon Redshift, and Amazon S3 and accommodates data processing use cases. Several use cases, such as building event-driven ETL pipelines, creating a unified catalog to search for data across multiple data stores, automatically generating the code for the ETL jobs, and building materialized views are a few examples of such.

Amazon Kinesis Data Firehose

Amazon Kinesis Data Firehose is a real-time streaming data service fully managed and delivered to destinations such as Amazon S3, Amazon Redshift, and Amazon OpenSearch service. This service aims to address the challenges around storing and analyzing high volumes of data along with a rich set of integration services. The built-in error handling, transformation, conversion, aggregation, and compression functionalities enable you to handle all the complexities that come along your way while trying to gain deeper insights into your data. You can easily collect the data, process it, analyze it in real-time, and stream through the pipelines, enabling monitoring and minimum maintenance through deep integration with other AWS services.

Features

Part of the Kinesis data platform, the Amazon Kinesis Data Firehose enables streaming data loading to data stores and analytics tools. Let us take a look at its key features to learn about its capabilities:

- **Real-time data loads**: You can configure how quickly you want to upload data and also use common compression algorithms such as Gzip and Hadoop-compatible Snappy. Kinesis Data Firehose supports batching and compressing the data, thereby allowing you to quickly receive new data.

- **Support for columnar data formats**: Apache Parquet, and **Optimized Row Columnar** (**ORC**) are all supported data formats for cost-effective storage and analytics. You get to integrate with services such as Amazon Athena, Redshift, EMR, and other big data-based tools.

- **Dynamic data partitioning**: You can define keys such as `customer_id` or `transaction_id` and deliver them to Amazon S3.

- **Data transformations**: You can integrate services such as AWS Lambda to prepare your data before loading it to the data stores. Pre-built Lambda blueprints are available to capture data from data sources such as Apache logs or system logs and convert it to JSON/CSV formats.

- **Support for multiple data destinations**: This is an important feature, especially for data coming in from a variety of data destinations, such as Amazon S3, Amazon Redshift, Amazon OpenSearch service, and HTTP endpoints.

- **Encryption support**: Your data can be automatically encrypted once it is uploaded to the destination endpoint, and AWS **Key Management Service** (**KMS**) enables the encryption for you.

- **Monitoring capabilities**: Several metrics are exposed via Amazon CloudWatch to monitor the health of your data streams and set alarms to ensure a successful data ingestion and delivery to the intended destination.

Use cases

Firehose's ease of use enables many use cases, such as the following:

- **IoT analytics**: Firehose provides the capability to continuously capture data from sources such as sensors. This data can be loaded into Amazon S3 and Amazon Redshift, which gives you near real-time access to metrics and insights via dashboards.

- **Streaming**: Kinesis Data Firehose can be used as a streaming tool where companies of any size can integrate their data across a wide variety of platforms. You can process terabytes of data in real time and apply ML models to the content in real time to enrich it with metadata and metrics.

- **Log analytics**: Firehose can record data from your applications to help you monitor and troubleshoot any issues more quickly than the typical triaging. Identifying the root cause and the ability to accurately detect errors within applications can be achieved through Kinesis Data Firehose.

- **Security Monitoring**: Integration support to destinations such as Splunk, means you can capture and transmit traffic flow and monitor network security in real time in case of a potential threat.

Unified governance

It becomes very important to build capabilities such as authorizing, managing, and auditing the data within the modern analytics architecture. Building these capabilities to manage security, access control, and audit trails is a complex, time-consuming, and error-prone task. AWS Lake Formation allows you to centrally perform these tasks and gives you the governance for enterprise-wide data sharing.

AWS Lake Formation

We've learned about data lakes and how time-consuming it is to configure a centralized, curated, and secured repository. AWS Lake Formation is a service that is easy to set up a data lake. AWS Lake Formation simply crawls through the data sources and moves data to the Amazon S3 data lake.

Features

Here are the features of AWS Lake Formation:

- **Ability to import data**: Other AWS services and databases make it easy to read and load your data into the data lake. With AWS Lake Formation, you can create custom ETL jobs with AWS Glue. You can also load your data into the data lake with Amazon Kinesis or Amazon DynamoDB.

- **Ability to catalog and label your data**: This makes it easy for users to discover and search information across available datasets. Text-based search is also supported.

- **Ability to transform data**: Such as processing data formats, to maintain consistency. Transformation templates are available where data can be transformed using AWS Glue or Apache Spark.

- **Ability to clean and deduplicate data**: Using ML Transform by easily learning the patterns and finding duplicate records.

- **Support for partitions**: This improves performance and cost-effectiveness.

- **Support for security**: This restricts access to a combination of columns and rows. This is especially helpful if your data needs to be **personally identifiable information** (**PII**) compatible.

Use cases

AWS Lake Formation can be used for the following scenarios:

- Building data lakes quickly using Lake Formation blueprints with built-in ML.

- Security management to centrally define and manage security. Encryption capabilities and audit logs enforces businesses to show compliance.

Self-service access to data with metadata can help data owners easily label data and search from data catalogs.

Analytics case studies on AWS

Companies can leverage AWS Analytics and accelerate their journey to capture actions from insights from their data. In this section, we will cover some case studies on how some companies were able to build platforms using AWS analytics services and innovate.

Case study – data analytics

In this case study, we will explore how AWS services can help improve efficiencies for companies that rely on data warehousing.

The business challenge

Locale, a company connecting people with local businesses, has grown from one city to a multi-national presence spanning more than 30 countries. Locale's traffic increased exponentially to more than 25 million average unique visitors of the app per month, 75 million average unique mobile web visitors per month, and 85 million average unique desktop visitors per month. With more than 100 million local reviews, Locale expanded to mobile, creating performance and efficiency issues stressing the analytics infrastructure.

Locale used data warehouses from internal teams, sales, advertising, product management, and mobile. However, it did not meet the requirements when it came to performance issues.

Solution overview

Locale took this opportunity to redesign its system. It had already been using a range of AWS services and its teams were familiar with the ecosystem. Locale explored Amazon Redshift and recognized the value of its petabyte scalability and highly managed warehouse and **Amazon Elastic MapReduce** (**Amazon EMR**), which provides a managed Hadoop framework. This improves the data processing through the ability to distribute dynamically across scalable Amazon EC2 instances. Locale uses Amazon S3 to store daily logs and photos of businesses.

Locale stores 24 months' worth of advertising data in Amazon Redshift. The ads team uses this data in different ways to train models that will result in making future ads more relevant. The company generates data of sizes in the range of multiple terabytes of logs daily. The data lands in Amazon S3. Locale created a Python package that integrates with Amazon EMR and writes and runs Hadoop streaming jobs as the data gets loaded into Amazon Redshift.

In addition to log processing, Locale used Amazon EMR to crawl the web and analyze the partner feed efficiently and cost-effectively. Locale created Redshift clusters on demand for specific analytics tasks. Developers were able to use dedicated resources for specific analytics tasks, and the elasticity helped Locale to answer ad hoc analytics queries immediately.

Locale used Amazon Redshift and Amazon EMR, which resulted in dramatic improvements in Locale's ability to access and analyze data. The teams owned Redshift clusters, creating a flexible infrastructure to meet the growing analytics needs of the company. The company was able to scale more effectively and cut query performance speeds from hours to milliseconds.

Case study – big data and data lake

Let us take a look at some of the company's data architecture and how they used AWS services to create centralized map data and keep up with their growing demand and innovate in terms of efficiencies.

The business challenge

A professional web hosting company, Hosting101, empowers entrepreneurs to build professional websites. The company began a large infrastructure and a shared Hadoop cluster on-premises. Various teams create and share datasets for collaboration. As the teams grew, copies of data started to grow in the **Hadoop Distributed File System** (**HDFS**). Many teams started to manage this challenge independently, resulting in duplication of efforts. This also made it very important for the data to be discoverable across a growing number of data catalogs and systems, which is a big challenge. The cost of storing several copies of the same data assets started to increase rapidly.

Data mesh solution overview

Hosting101 worked on building a data mesh based on a *hub and spoke model*. They wanted to centralize the data within the company, and the hub and spoke model enabled them to do so. The company's data platform team provided APIs, **software development kits** (**SDKs**), and Airflow operators as components to enable interaction with the catalog for various teams. Activities such as CRUD operations on the catalog meta store for products were centralized. Additionally, Hosting101 added a data governance layer to enable best practices for building data products across the company. This helped support data engineers and business analysts to create well-curated data products that are intuitive and easy to understand.

The data mesh enabled the teams to manage multiple petabytes of data across hundreds of accounts and enabled decentralized ownership of well-defined datasets with automation in place. Amazon S3 was used as the location to store, process, and create data partitions depending on the business needs.

AWS Lake Formation was used as part of the ecosystem for data producers to create new snapshots of the data every 15 minutes a day. AWS analytics services such as Amazon Athena are used to query the datasets. Managing permissions using Lake Formation and notifying the subscribed customer account via **Amazon Simple Notification Service (Amazon SNS)** enabled centralized governance.

With the move to AWS, Hosting101 successfully built a modern data platform and helped increase their velocity of building data products. They were able to transition from a monolithic platform to a model where data ownership has been decentralized.

Summary

In this chapter, we learned about data modernization and how it can help achieve efficiency and scalability on the cloud. We explored analytics, data lakes, and database offerings that AWS provides to deliver better, faster, and actionable insights out of the vast data that several companies have. As we build the right foundation for your data and leverage the right services, we can accelerate the process of data into action and develop state-of-the-art applications that can help accelerate your business goals.

In the next chapter, we will cover security on the cloud and how we can improve the core ability to protect data, identities, and applications running on the cloud as we consider compliance requirements, such as data locality, protection, and confidentiality with AWS's comprehensive set of services and features.

Part 3:
Security and Networking Transformation

In this part, we will introduce cloud security and how AWS can help you to leverage a secure cloud computing environment for your applications. You will learn about AWS's Shared Responsibility Model and how you can benefit from the cloud to build and run your applications on a platform designed for high security without the capital outlay and operational overhead of a traditional data center. We will deep-dive into some of the important service offerings you can leverage to run your applications while meeting your security objectives. We will review the available AWS controls for network security, configuration management, access control, and data security. This part also covers networking on the cloud and the broadest and deepest set of networking constructs that AWS offers. You will learn about AWS infrastructure and how to maintain the highest availability levels for your mission-critical workloads using AWS Regions and the Availability Zone model.

This part comprises the following chapters:

- *Chapter 10, Transforming Security on the Cloud Using AWS*
- *Chapter 11, Transforming Networking on the Cloud Using AWS*

10
Transforming Security on the Cloud Using AWS

With digital transformation on the radar of every company across any industry to drive opportunities, one core aspect that often gets neglected is security. Security can be the top weakest link of your business if not well planned. Businesses that are transforming to capture new user behaviors, and change their business models as a result, often depend on multiple aspects. Aspects such as technology, processes, and people changes can result in security flaws and risks if not properly planned. Building an adequate security strategy to protect your digital transformation remains one of the biggest tasks for executives.

Security is of paramount importance for any business to protect its applications from deliberate theft, leakage, integrity compromise, and deletion. Discovering flaws and vulnerabilities and taking remediation actions takes time and maturity. The **Chief Information Security Officers (CSOs/CISOs)**, SecOps team, developers, and operations team members are usually charged with the task of solving the security puzzle. With an increasingly complex cyber threat landscape, many organizations look for recommendations to bolster their security operations on the cloud. At a high level, the fundamentals of securing workloads on cloud and on-premises networks are similar when it comes to protecting the assets, detecting malicious activity, responding to security events, and recovering afterward. However, there are some unique aspects while operating in a cloud environment. In this chapter, we will cover the following main topics that will give you in-depth knowledge of how to make security an enabler of your digital transformation:

- Understanding the security implications of digital transformation
- Introduction to security on the cloud
- Identity and Access Management (IAM) using AWS
- Fraud and anomaly detection using AWS
- Network and application protection using AWS
- Data protection using AWS

Understanding the security implications of digital transformation

There is a large uptick in security vulnerabilities for enterprises that do not follow a security-by-design methodology in their digital transformation efforts. Enterprises often find themselves dealing with security vulnerabilities in their infrastructure, web applications, **Application Programming Interfaces** (**APIs**), and microservices that can lead to business and reputation loss.

Legacy vulnerabilities and misconfigurations remain a challenge for many enterprises and can be used as an entry point for enterprise value chain attacks. Let us learn about the main security challenges that can introduce cybersecurity risks in today's digital world:

- **Unsecured data**

  With the increase in the number of web transactions, the volume of data is increasing and is vulnerable to attacks. This data usually flows through three distinct areas:

  - Data at rest within applications, whether it is on-premises, cloud, or hybrid
  - Data in transit.
  - Data at endpoints such as data residing in user devices.

  Data breaches are possible across all these areas where hackers can gain access and create a ripple effect on the dependent systems.

- **Unsecured systems**

  As organizations move to cloud-native architecture, they may still be in the process of selecting the right security tools and mechanisms. This opens up access for hackers to exploit the systems and devise security breaches. In addition, connection to the internet, while businesses expand the user base, can become an ever-growing attack surface area for cybercriminals. When critical systems are compromised, high-profile ransomware attacks often become news headlines, which is detrimental to any business.

- **Unsecured Software-as-a-Service (SaaS) or third-party tool providers**

  Many companies depend on software services offered by SaaS or third-party providers. Security is an important pillar for SaaS providers to offer reliable software. On the surface area, the third-party providers and SaaS vendors guarantee capabilities of securing information but are often secretive about how they do it. Malicious attacks through SaaS applications can result in exposing organizations to malware, phishing, ransomware, and insider threats.

  Not having an understanding of data privacy and vendor liabilities is worrisome. A lack of transparency leaves users and customers uncertain of the kinds of regulations that are in place.

- **Unsecured storage**

  When you store data in physical drives or on-premises servers, having physical access to the servers will put your data at risk. When you have data to manage and opt for APIs and storage gateways for easy migration that are insecure, it can lead to data loss. Unsecured storage is prone to phishing attacks, leading to business breaches. Third-party storage providers encourage companies to **Bring Your Own Device** (**BYOD**), which can pose a high security risk when devices are compromised and act as an entry point for cyberattacks.

- **Unsecured networks**

  The attack surface dramatically increases with the increase in the number of remote endpoints that users can use, such as personal phones and laptops. Increasing the attack surfaces means creating security inconsistencies where there is no control over personal devices that the users use.

- **Challenges with staffing skills**

  There is a skill shortage when it comes to organizations having cybersecurity professionals implementing preventive protection strategies and building risk-free environments. The 2022 Cybersecurity Workforce study (`https://www.isc2.org/Research/Workforce-Study`) calls out that there is a requirement to grow cybersecurity by 65% to effectively defend organizations' critical assets. Security is a nuanced skill set, and organizations should invest in staff education initiatives to keep their employees informed on topics such as social engineering tactics, clean desk policies, and acceptable usage of work devices and the network.

- **Challenges with compliance**

  The usage of cloud computing services adds a new dimension to regulatory and compliance requirements such as the **Health Insurance Portability and Accountability Act** (**HIPAA**), **Payment Card Industry** (**PCI**), and more. Eliminating blind spots and assessing the security posture can ensure that you are preventing external and internal cybersecurity threats. You will need to include the cloud provider infrastructure, interfacing APIs, and other details in compliance and risk management processes. If you don't have an IT team to ensure proper data access control or configurations, it can easily escalate to a compliance breach, putting your organization at risk of penalties and lawsuits.

While many companies want to improve their products by using cutting-edge technologies, security cannot be overlooked. In the next section, we will learn how moving to the cloud changes the dynamics of how you should implement security while running workloads on the cloud.

Introducing security on the cloud

As enterprises embrace cloud computing, new challenges arise, and it is important to have security controls, technologies, practices, and procedures to protect the data on the cloud. New security challenges often come up during cloud migration, and you have to balance the security posture of your organization. In order to navigate through the complexities of compliance, choosing a certified compliance cloud service provider and ensuring proper data access control is in place are essential components. Cloud security is dedicated to securing your workloads on the cloud, including data privacy across applications and platforms; let us take a look at what cloud security is in the following note. Cloud providers offer a range of security services and features to secure your assets. These collections of security measures can be used to protect infrastructure, applications, and data.

> **Cloud Security**
>
> A collection of processes, mechanisms, protocols, best practices, and technology to protect the applications running on the cloud, secure data, and address modern security threats or challenges.

Traditional IT security is different from cloud security. Cloud security comes with modernized security protocols to mitigate the security challenges and helps in staying compliant. Cloud security protocols are dedicated to securing cloud computing systems that span across infrastructure, applications, and platforms. When it comes to the scope, the following are the aspects that need to be taken into consideration when building a strategy for cloud security:

- Data storage such as hard drives.

- Physical networks and servers such as routers, power, and cabling.

- Virtualization frameworks such as host machines, virtual machine software, and guest machines.

- Middleware such as API management.

- Runtime environments such as execution and the upkeep of running programs.

- Hardware devices such as computers and **Internet of Things** (**IoT**) devices.

- Operating systems to house the software running on hardware devices.

- Applications running on the cloud, including software, productivity suites, and more.

While we will discuss in detail who has the ownership of the preceding components in the *Shared responsibility model* section, it is also important to understand how these can be categorized. Let us look into the cloud security categorizations.

Data security

Data security is the discipline of protecting the data of cloud-based applications, systems, and their associated users. The core aspects of information security, such as who has access to that data, the role of the users in accessing the data, and the location of the data, are all essential when implementing data security on the cloud. Tenets of data security and governance that apply to the cloud include data confidentiality, integrity, accessibility, and availability. These tenets are relevant regardless of the cloud model (public, private, hybrid, or community cloud) and cloud computing category (SaaS, **Platform-as-a-Service** (**PaaS**), **Infrastructure-as-a-Service** (**IaaS**), or **Function-as-a-Service** (**FaaS**)). Data security needs to be considered across all stages of the data life cycle, from development and deployment to the management of the cloud applications and infrastructure. Protecting data in transit or data at rest or introducing a layer of opacity between your applications are all additional protection controls that you can impose.

Identity and access management (IAM)

IAM is the security discipline to validate who a user is and what specific permissions they have. Enterprises find it difficult to set up their own IAM solutions because IAM is considered an indirect value addition to security and doesn't directly add to profitability. However, IAM benefits your organization's security posture in many ways, such as the following:

- The ability to initiate, capture, and manage user identities and access permissions to the cloud resources

- The ability to authorize and evaluate roles and permissions based on employees' needs and tasks

- The ability to protect sensitive data and applications using just one set of credentials through AWS Identity Center (formerly **Single Sign-On (SSO)**)

- The ability to maintain the compliance of processes and protocols based on regulatory requirements

- The ability to automate security access to your network more efficiently by decreasing the manual effort, time, and resources

Identifying all the business needs to map with the specific areas of IAM policies will impact your enterprise's security functions. Businesses will get the most out of IAM by making sure to centralize security and critical systems around identity. As an IT professional, having an understanding of IAM-specific and broader security functions will help you ward off any threats and assess the organization's security posture in an efficient manner.

Compliance

Cloud compliance is the principle that systems running on the cloud must adhere to the standards defined in compliance regulations such as HIPAA, PCI, **General Data Protection Regulation (GDPR)**, **National Institute of Standards and Technology (NIST)**, and more. In traditional on-premises data centers, you are responsible for your entire infrastructure and network. However, once your applications are running on the cloud, you should be aware of the cloud provider's compliance with the general data protection laws, HIPAA in the United States, GDPR in the EU, and the Privacy Act 1988 in Australia. This should be one of the first aspects in your discussions when choosing a cloud provider. Protecting user privacy as set by the legislative bodies is important for organizations to abide by their compliance requirements.

Governance

Governance is a discipline that focuses on threat protection, detection, and mitigation to guard your systems on the cloud. Organizations should be equipped with the required security and governance controls to be able to respond to threats. When you move to the cloud, building a cloud security governance model relevant to your industry is critically important. Understanding what governance is when it comes to applications running on the cloud and risks to manage and reacting to the changing world to maintain consistency, scalability, and security are where a governance model helps.

Re-evaluating the cloud security for your data and applications and protecting it requires adjusting the existing IT practices and educating your teams to be well versed in cloud security responsibilities. In the next section, we will learn about these responsibilities and the shared duties of maintaining and securing infrastructure and systems on the cloud.

Shared responsibility model

The shared responsibility model is a cloud security framework that provides guidance on the roles and responsibilities of cloud service providers and consumers. The model typically varies by cloud service provider, where each party maintains complete control over the resources, processes, and functions. This model is targeted to relieve the operational burden of maintaining and controlling the physical security of the facilities that physically host the cloud services.

For AWS, security is a divided responsibility between AWS and the user, as shown in the following diagram. It is important to develop a solid understanding of this model when you move to the cloud and continue to operate on it. When you move your application assets and data to the cloud, security responsibilities become shared between you and your cloud service provider:

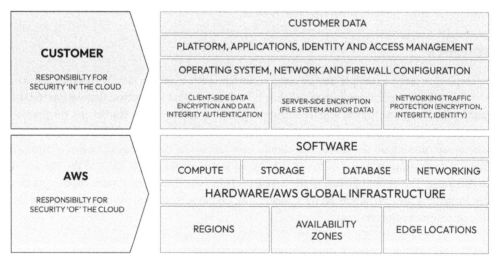

Figure 10.1 – Shared responsibility model on AWS

Responsibilities owned by the cloud service provider (AWS)

Cloud service providers are responsible for the security *of* the cloud. AWS is responsible for operating, managing, and controlling the components from the host operating system, the virtualization layer, and the physical security of the facilities where the infrastructure is hosted. The infrastructure typically comprises hardware, software, and networking components, including the facilities that host the cloud services.

This is a crucial benefit for cloud consumers, where they are relieved from the burden of managing the security of the physical infrastructure. The end users can instead use the AWS control and compliance documentation for control evaluation and verification procedures. You do not have to worry about the AWS data centers or setting up any security tasks for physical data center protection.

Responsibilities owned by you (the end user)

The end user is responsible for security *in* the cloud. As a consumer, you need to take the required steps to protect the platform, data, applications, operating system, network, and configurations. The cloud consumer's responsibility varies based on the specific AWS cloud services they select to use.

For IaaS services such as Amazon **Elastic Compute Cloud** (**EC2**), it is recommended to configure all the necessary security controls and management tasks. Tasks such as applying security patches on the operating system and configuring AWS-provided firewalls on the instances come under this category. Patch management, configuration management, and protecting service and communications are the responsibility of the consumer.

For services such as Amazon **Simple Storage Service** (**S3**) and Amazon DynamoDB, you manage your data, including encrypting it, using AWS-provided tools. Evaluating each use case and understanding how to apply the necessary security controls are critical aspects of applying security practices on AWS.

Top cloud security considerations

Securing your cloud workloads is job zero. Operating in the cloud without a proper security posture can lead to many disastrous consequences as we discussed earlier. Successful cloud security implementations can be achieved through zero-trust principles. The goal is to have consistent security across physical and virtualized resources and protect your applications from threats, whether it is through the cloud environment or physical network.

Including stakeholders such as the CISO, **Cybersecurity Compliance Officer** (**CCO**), audit teams, security architects, and engineers is also a key aspect.

The following are the five key considerations for implementing successful cloud security operations.

Identify and alert on security weaknesses

Knowing and understanding your cloud service provider's security foothold is the first step to understanding the effective mechanisms for running workloads on the cloud. Cloud service providers typically partner with third-party vendors or independently audit cloud security to boost their security posture. The best practices and solutions published by these cloud service providers are typically a good fit for organizations of any segment. The important aspect is to understand and identify the gaps where hackers can gain access and proactively address those issues. Examining these blind spots will help boost your cloud security overall.

Configuring the tools for cloud security monitoring is a recommended practice of using automated solutions to identify potential threats to cloud resources. This enables security teams to identify patterns from malicious activities and take necessary preventive actions to ensure that the applications are running smoothly. There are several risks that you can monitor. Misconfigurations, data loss, API vulnerabilities, and compliance violations are some of the risks that can result in extreme damage and should be monitored. Amazon GuardDuty and AWS Security Hub provide continuous visibility and detection of threats for AWS accounts and workloads.

Implement access control regulations

Giving access to your cloud environment across the organization should be designed to incorporate the least privilege policy. Keeping unauthorized users out becomes the core aspect of access control regulations. This includes identity management, authorization, and authentication configurations. Three capabilities that must be included in every organization's access management approach are the following:

- The ability to identify and authenticate users

- The ability to assign the right access permissions to the users

- The ability to develop and implement access control policies for your cloud resources

Least privilege access is key to achieving a cloud environment resilient to security threats. By automating the preceding actions, you will be able to meet your organization's highly dynamic and growing business functions.

Encrypt your data

The ultimate goal of securing your workloads is to prevent any unauthorized access to your data and to keep it protected. Encrypting your data on the cloud ensures that you are aligned to achieve this goal. There are three aspects involved in data protection:

- Protecting data from unauthorized access

- Ensuring data is accessible during failures or errors

- Preventing accidental data disclosure

In order to maintain visibility and control over data, it is important to understand how it is being accessed and used in your organization. Classify your data based on the nature of usage, its criticality, and its sensitivity, to apply the appropriate data protection controls. Encrypting data at rest and in transit is the best way to protect your data.

Secure your applications

Monitoring cloud-deployed applications by using the provided services is your responsibility. You need to design and implement monitoring carefully, ensuring the end-to-end systems are completely integrated with monitoring capabilities. With the growing application development and deployment on the cloud, the development and operations teams (DevOps) are at a unique juncture to apply security in both breadth and depth. It is important for DevOps to own the security aspect; hence you see the term **DevSecOps** becoming more prevalent now. This discipline is foundational to integrating security objectives in the early life cycle of software development.

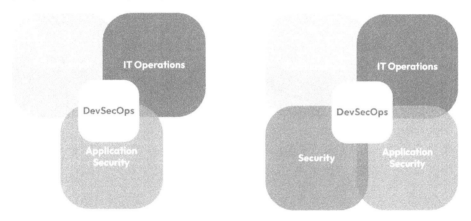

Figure 10.2 – DevOps versus DevSecOps

As the preceding diagram suggests, the difference between DevOps and DevSecOps is the different goals that each of these practices offers. DevOps focuses on efficiency, while DevSecOps focuses on security. DevOps breaks down the boundaries between software development and operations to be more agile. This collaboration is taken one step further with DevSecOps to add security integration at every step of the process. Previously, security was bundled as a loose collection of practices incorporated into the development and operational blueprints. DevSecOps is the natural evolution of DevOps, where the security practices are applied as an extension of the activities and injected into the early life cycle of application development.

Bringing security into the DevOps fold enables changes in the culture of the development teams where they keep security in mind while they build applications. Greater collaboration between the development and security practitioners is cultivated by achieving faster delivery with an improved security posture.

Integrating security into the **Continuous Integration/Continuous Delivery (CI/CD)** pipeline enables an excellent foundation for automated security testing and validation without requiring manual intervention:

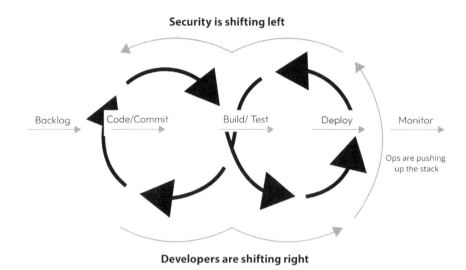

Figure 10.3 – DevSecOps adoption

As shown in the preceding diagram, security practices are integrated during the initial stages of application deployment so that threat modeling, static analysis, and policy engine checks can be incorporated anytime a developer checks in their code. This helps in confirming that the open source dependencies and compatible licenses are free of any vulnerabilities. Security integration tests are employed to generate automatic feedback enabling quick iteration and triaging of security issues. By implementing these actions in an automated manner, your organizations are integrating the security objectives at every phase of application development and removing the burden of gatekeeping the additional processes.

Continuously train your enterprise

IT staff that have cloud security skills should be a top priority in the cloud. To protect an enterprise and its data, it is important to provide and support continuous education on processes and frameworks. Cloud certifications such as **Certified Cloud Security Professional (CCSP)** (`https://www.isc2.org/Certifications/CCSP`), **Certificate of Cloud Security Knowledge (CCSK)** (`https://cloudsecurityalliance.org/education/ccsk/`), **GIAC Cloud Security**

Automation (**GCSA**) (`https://www.giac.org/certification/cloud-security-automation-gcsa`), and AWS Certified Security – Specialty (`https://aws.amazon.com/certification/certified-security-specialty/`) provide security best practices to stay ahead of ever-changing security threats.

Architecting secure workloads on the cloud with AWS

When it comes to protecting data, systems, and assets running on AWS, there are AWS whitepapers that you can take advantage of and use to build your workload security. There are basic tenets that can be employed, summarized as follows:

- **Have a strong identity foundation**: The **Principle of Least Privilege** (**POLP**) is a fundamental point that is more important than ever before when it comes to securing your resources. POLP enables users, systems, and processes to have access to the resources that are absolutely necessary to execute specific actions. This helps enforce the separation of duties for each transaction within the resources and eliminate most modern attacks. By limiting privileges, you can minimize the overall attack surface of your organization and reduce the pathways a bad actor can exploit the resources that they have access to. By limiting the superuser and administrator privileges, you can prevent, detect, and fend off malicious activities.

- **Implement traceability**: When you are monitoring and alerting on any changes made to your environment in real time, your ability to track down the root causes of malicious activities becomes straightforward. Integrating logging with your systems and audit actions automation can instantaneously help remediate any compromises.

- **Apply security at all layers**: The **Defense in Depth** (**DiD**) approach is the process of implementing security mechanisms and controls to protect the confidentiality, integrity, and availability of your workloads. This is an intentionally layered approach to ensure that the security controls are baked into layers (edge of the network, **Virtual Private Cloud** (**VPC**), load balancing, compute instances, operating system, application, and code) and bolsters security against many attack vectors.

- **Automate security wherever possible**: Security automation is expanding its scope and evolving rapidly as IT businesses start to witness its potential in catering to implementing security at the cloud scale. Security automation refers to the automation of the detection and prevention of cyber threats. It includes implementing overall threat intelligence and defending against future attacks by your SecOps team executing security best practices. And not to forget streamlining communication, such as alerting the key stakeholders of the business and mitigating the risk.

- **Protect your data**: Data protection is the process of protecting sensitive data from damage, loss, or corruption. The idea is to ensure data stays safe and remains available to the users at all times. Classify your data and use controls to encrypt and protect your data.

Data in transit, or data that is moving from one place to another over the internet/private network, is easily susceptible to attacks when it is on the move. In order to secure the data in transit against malware attacks or intrusions, it is recommended to apply the required network security controls such as firewalls and network access controls.

Data at rest is data that does not move from network to network or device to device. Data at rest is often vulnerable to intruders within a given network. In order to reduce the risk of data breaches, use mechanisms and tools to reduce or eliminate the need for direct access.

- **Prepare for security events**: Security events such as data breaches, ransomware, or a cyberattack, be it malware, **Denial of Service** (**DoS**), SQL injection, or zero-day exploit, can compromise any routine business operations. Preparing to respond to such incidents by having a thorough incident management workflow followed by investigation mechanisms that align with your organization's protocols is important.

There are several AWS resources that offer security guidelines in addition to those previously mentioned:

- **AWS Cloud Adoption Framework (CAF)**

  CAF utilizes years of experience and best practices to help you to transform and accelerate your business outcomes while using AWS services. The **security perspective** of CAF enables you to achieve capabilities such as security governance, threat detection, data protection, security assurance, vulnerability management, application security, IAM, infrastructure protection, and incident response.

- **AWS Well-Architected Framework**

  The Well-Architected Framework enables cloud architects to build and run their applications and workloads on the cloud around six pillars: operational excellence, security, reliability, performance efficiency, cost optimization, and sustainability. The security pillar provides in-depth guidance, best practices, and current recommendations that can be used during the design, delivery, and maintenance of AWS workloads. This helps in improving the security posture overall so you can operate your workloads securely on the cloud.

Building a threat model

The security pillar of the framework calls out threat modeling as a specific best practice for building a good foundation of secure workloads. As the complexity of IT systems is growing, integrating threat modeling into your application development life cycle becomes a crucial component to mitigate potential security threats bringing down your business value. The number of use cases adds to the complexity, making it ineffective to use ad hoc approaches to find and mitigate threats.

Threat modeling offers an approach that is systematic to capture the potential threats to a workload and implement mitigations to make sure the resources of your organization are not impacting your security posture. As you are equipped with the right resources to find and address issues early in the design process, the mitigations will result in a low cost when compared to doing the same later in the application development life cycle. Let us look at the core steps of threat modeling at a broader level:

1. Define security requirements.

2. Analyze every component in the design diagram and identify entry points for potential threats.

3. Identify a list of threats and rank them.

4. Determine countermeasures and mitigation per threat.

5. Create and review the risk matrix to check whether the threat is fully mitigated.

More details on threat modeling can be found in the SAFECode Tactical Threat Modeling whitepaper (`https://safecode.org/wp-content/uploads/2017/05/SAFECode_TM_Whitepaper.pdf`) and the Open Web Application Security Project (OWASP) Threat Modeling Cheat Sheet (`https://cheatsheetseries.owasp.org/cheatsheets/Threat_Modeling_Cheat_Sheet.html`). These references can help you guide in identifying and communicating information about the threats that can impact a particular system.

- **AWS Marketplace**

    The Marketplace lists software from high-quality and vetted sellers to maintain the security posture. Alternatively, you can also connect with IQ experts that are certified freelancers and consultants. A list of security offerings is available based on your use cases where you can pay directly through AWS after the work is completed and satisfied. Example use cases are OpenVPN on AWS, securing your WordPress site through **Transport Layer Security** (**TLS**), AWS Security Hub automated response and remediation, and so on. This helps in accelerating your work and not having to build homegrown solutions.

- **AWS Security Competency Partners**

    The offerings that are available through the partners provide deep technical expertise and proven success in delivering security-focused solutions suitable for workloads. You can easily search, buy, deploy, and manage solutions that are cloud-ready in a matter of a few minutes. Security categories such as network and infrastructure, host and endpoint, data protection and encryption, governance, and compliance are available for you to take advantage of the SaaS products.

- **Cloud Audit Academy (CAA)**

    The CAA is a security audit learning path designed by AWS. The curriculum dives into global industry standards for cloud-specific auditing, risk, and compliance processes. Education on recognized domains such as data security and privacy, logging and monitoring, configuration management, and several others is covered.

- **AWS Prescriptive Guidance**

  AWS publishes time-tested strategies, guides, and patterns within AWS Prescriptive Guidance to help with cloud migration and modernization efforts. You can find strategies, guides, and tools for several security-based use cases, such as taking backups securely in AWS, playbooks for large migrations, automatically encrypting **Elastic Block Storage** (**EBS**) volumes, and many more.

- **AWS Security Reference Architecture (SRA)**

  SRA provides a holistic set of guidelines to design, implement, and manage AWS security services and align with the best practices. The architecture is built around a three-tier web architecture similar to real-world applications comprising web applications and data tiers. The architectural guidance goes deeper than the fundamentals and touches on security principles, examples, and categories.

In addition to this, AWS launches blog posts, security bulletins, and many other reference materials so that security teams can stay up to date with the latest offerings and best practices.

Identity and access management using AWS

AWS IAM securely controls access to your resources on AWS. Among various security services that AWS offers to operate your applications on a secure platform, IAM is a critical service. When you deploy applications on AWS, IAM can be used to create, manage, and control access to the AWS resources shown in the following diagram:

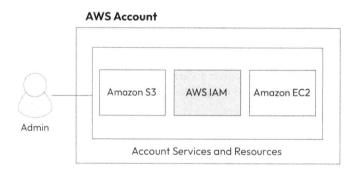

Figure 10.4 – AWS IAM

With IAM, you get to provide the following capabilities centrally within your organization:

- **Manage users and their access**: You can create IAM users, assign credentials, and manage the permissions to specific operations that a user can perform.

- **Manage roles and their permissions**: You can create IAM roles and manage permissions to specific operations. Operations include any action that can be performed by the users/entity/ AWS service that assumes the role.

- **Manage federated users and their permissions**: You can enable identity federation and allow the authorized users/groups/roles in your company to access the AWS resources.

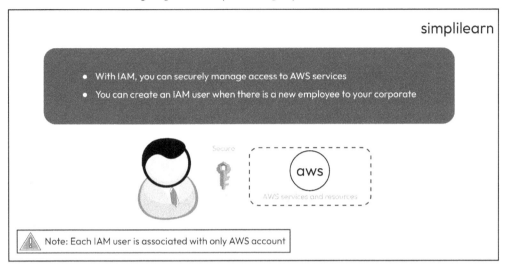

Figure 10.5 – IAM capabilities

Without IAM, corporate environments need to create their own framework and have a centralized IT office to manage the authentication and authorization of resources. This often involves manual processes where passwords are shared across the corporate offices, and you need to call an admin to reset passwords over the phone.

IAM can be leveraged by the IT administrators of your organization to manage user identities and the granular level of permissions to the AWS resources. As a result, you can establish a secure and efficient approach while connecting to AWS resources.

Understanding the concept behind IAM

IAM follows POLP, meaning that users are given the least amount of access and responsibility required to complete their tasks. As shown in *Figure 10.6*, applying POLP is a fundamental concept within IAM:

What Is Principle of Least Privilege?

 The "Principle of Least Privilege" (POLP) states a given user account should have the exact access rights necessary to execute their role's responsibilities—no more, no less. POLP is a fundamental concept within identity and access management (IAM).

POLP Best Practices:

- Make least privilege model the default for all accounts.
- Elevate privileges on a situational and timed basis only .
- Monitor and track all network activity.
- Adopt a flexible access managemnt platform so that privileged credentials can be securely elevated and easily downgraded.
- Identify and separate high-level from lower-level system functions.
- Audit privileges granted to users and applications regularly.

POLP Benefits:

- ✓ Creates an environment with fewer liabilities.
- ✓ Limits the possibility of data breaches.
- ✓ Protects against common attacks, like SQL injections.
- ✓ Data classification promotes a healthy network.
- ✓ Superior data security and audit capabilities.

Figure 10.6 – POLP

A simplified IAM workflow has the following seven components:

- **Principal**: This is an entity that performs actions on an AWS resource. The principal can be a user, a role, or an application.

- **Authentication**: This is the process of validating the identity of the principal that is accessing the AWS resource. Valid credentials or keys must be provided to complete the authentication process.

- **Request**: This is sent by a principal with details such as the specific action on the AWS resource and the resource to execute the action on.

- **Authorization**: Using POLP, all the resources are denied. IAM is used to authorize a request based on the matching policy. AWS approves the action after successful authentication and authorization.

- **Actions**: These typically view, create, edit, or delete an AWS resource.

- **Conditions**: Although optional, these allow you to specify conditions for when a policy is in effect. You can build expressions to use operators such as equal, less than, and more than. This is used to match the condition keys and values in the policy.

- **Resources**: This is a set of AWS resources specific to your AWS account.

Figure 10.7 – AWS IAM

As shown in the preceding diagram, you can define who can access what resources by specifying fine-grained permissions. IAM then enforces the permissions for every request. IAM is a powerful service to maintain the security posture in the cloud. This is a free-of-charge service and helps with securing the cloud, which is your responsibility, so it is important to understand its usage and concepts.

Features

The following list includes the main IAM features:

- **Fine-grained access control**: With IAM, you can specify and control access to the AWS resources. IAM policies let you grant permissions to IAM roles.

- **Multi-factor authentication (MFA)**: MFA is an AWS best practice that provides a second authentication factor in addition to the sign-in credentials. You can enable MFA at the AWS account level.

- **Identity federation**: Identity federation is a process of trust between two parties. This helps manage users centrally within a standard identity provider and govern their access to multiple applications and services. You can define federated access permissions based on their roles in a single centralized directory. When you use multiple directories, you can consider IAM as your design alternative.

- **IAM Access Analyzer**: IAM Access Analyzer can help streamline permissions management throughout each step of a cycle of granting the right fine-grained permissions.

Business use cases

With IAM, you can practice managing digital identities throughout their life cycle under a centralized management system. Let us look at some of the key use cases.

AWS Identity Center (successor to AWS Single Sign-On)

You can manage sign-in security for your workforce identities and workforce users to manage access across all their AWS accounts and applications. For many enterprises, implementing SSO using their own technologies is a daunting task. The benefits of using IAM are fast setup and reducing the pain of maintaining several user identities and passwords. You can simply assign employees roles in the enterprise directory and connect to the management or member account of AWS Organizations. You can also categorize users based on their job functions. For example, you can grant database administrators a broader set of permissions to Amazon **Relational Database Service** (**RDS**) in staging accounts and limit their access to the production accounts. With IAM, the administration of access permissions is simplified, and you can assign access to groups directly instead of individual users. There are many popular SSO software products, such as Okta Workforce Identity, Rippling, JumpCloud, and LastPass, to sign in to multiple applications or databases with a single set of credentials. To qualify products, your strategy should ideally incorporate the following:

- Have one portal that allows users to access applications or databases.

- Automate authentication wherever possible so the users are not logging in multiple times.

- Centralize authentication across multiple applications.

- Access to applications and data should be secured.

- Integrate login access to business applications.

The goal of SSO products is to improve the ease of use and minimize the work for IT admins and developers by centralizing access management.

Establish permissions guardrails

Many enterprises of any size require that their workloads and data are PCI DSS or HIPAA compliant. For this, having an enterprise-wide strategy for access management guardrails is critical. IAM Access Advisor can be used to restrict permissions to ensure that services that should not be accessed without having proper responsibilities are protected. IAM Access Advisor can be used to view the service-last-accessed information, and depending on the organization's activity, permission guardrails can be set so that the access is restricted to the services that are required to run existing workloads.

Centralized provisioning/deprovisioning

Many enterprises want to allow their engineering teams to experiment with AWS services in lower or sandbox environments as they move toward production-ready applications. These teams need access to various sets of AWS services and resources, ensuring POLP is honored. For example, application teams generally should not have administrator access to take periodic Amazon EBS snapshot backups or Amazon CloudWatch Events rules to send events to centralized information security accounts.

IAM can be used to create a centralized and automated workflow to help with the creation and validation of IAM policies for such teams working in various environments. Your security professionals can customize the workflow to align with the specific requirements of the security team.

Fraud and anomaly detection using AWS

Enterprises have the majority of their businesses running online and have to be constantly on the lookout for fraudulent activities such as payments with stolen credit cards. AWS has a stack of services such as AWS Security Hub, Amazon GuardDuty, Amazon Inspector, AWS Config, AWS CloudTrail, and AWS IoT Device Defender to evaluate, record, track, automate, and protect resources on the cloud.

AWS Security Hub

AWS Security Hub gives you a consolidated view of your enterprise's security posture in AWS. This helps you to keep track of your environment and validate against the security standards and the latest AWS security recommendations. Security Hub overlaps with **Security Information and Event Management** (**SIEM**) tools, but it is doesn't replace them. There are benefits to using both Security Hub and SIEM together to get the best out of both tools. Doing so gives you a thorough view of security, operations, and compliance data.

Features

Some of the main features of Security Hub are as follows:

- The ability to combine the Security Hub findings across AWS Regions. You can then generate a single dashboard view.

- The ability to combine security and compliance data into a single centralized dashboard.

- The ability to build scorecards for security and compliance dimensions.

- The ability to create correlation using the security and compliance findings.

The searchable nature of the data helps you to achieve improved security posture across the organization. Some organizations find it very useful to aggregate these findings across multiple admin accounts and inject them into Amazon OpenSearch Service to be able to draw meaningful insights.

Amazon GuardDuty

GuardDuty is a continuous security monitoring service that is used to analyze data across Amazon VPC, **Domain Name System** (**DNS**), and other Amazon sources. This threat detection service helps you to guard against using compromised credentials, unusual activities through malicious IP addresses, and so on. GuardDuty uses various techniques such as machine learning, anomaly detection, and threat intelligence to identify potential breaches or threats.

With the machine learning model under the hood, GuardDuty operates on the continuous stream of API invocations based on the AWS CloudTrail activity on your AWS accounts. GuardDuty learns to distinguish normal activity versus suspicious behavior within your AWS accounts. It analyzes the security relevance of the user activity, taking context into consideration. The threat detections will help the DevSecOps teams to proactively identify any potential attacks and quickly trigger incident response workflows without losing any time for detections.

Features

Amazon GuardDuty can help with continuously monitoring and protecting your AWS accounts. All the insights are stored in Amazon S3. The billions of activity records produced by AWS CloudTrail management events, Amazon VPC Flow Logs, and DNS logs are analyzed, and any malicious activity that can impact your accounts is identified.

GuardDuty comes with many features, with some of the critical capabilities listed as follows.

Account-level threat detection

With GuardDuty, you can get access to built-in detection techniques. AWS maintains the detection algorithms and continuously improves on them. When it comes to detection categories, there are four of them that AWS supports:

- **Reconnaissance**: Unusual API activity, intra-VPC port scanning, unusual patterns of failed login requests, or unblocked port probing from a known bad IP are all activities under this category.

- **Instance compromise**: Activities such as cryptocurrency mining, backdoor **Command and Control (C&C)** activity, malware using **Domain Generation Algorithms (DGAs)**, and unusually high volumes of network traffic fall into this category.

- **Account compromise**: Activities include common patterns such as attempts to disable Amazon CloudTrail logging, changes that weaken the account password policy, infrastructure deployment in an unusual Region, and API calls from known malicious IP addresses.

- **Bucket compromise**: This includes an Amazon S3 bucket compromise due to suspicious data access patterns indicating credential misuse.

You can build in-house detection solutions that are customized to your business requirements and develop your own threat detection intelligence of known malicious IP addresses.

Continuous monitoring

With Amazon GuardDuty, you don't have to do the heavy lifting of continuously monitoring and analyzing your AWS accounts. The seamless integration with services such as Amazon CloudTrail, VPC Flow Logs, and DNS Logs does not require infrastructure. SecOps teams can focus on how to respond quickly to keep your organization secure at scale and focus on innovating.

Efficient prioritization

With Amazon GuardDuty, you get three severity levels (*Low*, *Medium*, and *High*) to prioritize these threats accordingly. A *Low* severity level indicates an audit before taking actions, such as blocking them, a *Medium* severity level indicates suspicious activity, and a *High* severity level indicates that the specific resource is compromised and is actively being used for unauthorized purposes.

A case study with Amazon GuardDuty and AWS Security Hub

In this case study, we will explore how companies can identify threats by continuously monitoring network activity and account behavior using AWS Security Services.

The business challenge

WagenAuto, a global automobile manufacturer, operates more than 100 production plants across 20 European countries. It creates products for several automobile brands and uses on-premises and cloud-based solutions. With the growing scale, the company was looking to strengthen its security posture by detecting vulnerabilities, having a centralized view, and responding to threats in a timely manner.

Solution overview

WagenAuto evaluated Amazon GuardDuty and decided to activate it across their 200 plus AWS accounts using AWS Organizations to centrally manage and govern their AWS resources that were growing. Enabling GuardDuty for all regions was an important step for the **Security Operations Center** (**SOC**) team to ensure full coverage. The company's SOC team managed a list of approved AWS Regions on which they can deploy their workloads. They enforced **Service Control Policies** (**SCPs**) across their AWS accounts. During the account provisioning process, the team created automation to apply SCPs and prevent individual accounts from modifying Amazon GuardDuty.

Additionally, the company integrated Amazon GuardDuty and AWS Security Hub to get a comprehensive view of the security alerts and maintain the posture for the SOC team. The SOC team was able to add rules/filters in Amazon CloudWatch to trigger a custom AWS Lambda function for different types of events. By connecting CloudWatch events from GuardDuty to Lambda functions, the SOC team was able to implement automated remediation actions for each type of GuardDuty finding.

Using Amazon GuardDuty, AWS Security Hub, and AWS Organizations, WagenAuto was able to automatically identify security threats and quickly take corrective actions to protect their AWS accounts, business applications, and infrastructure.

Network and application protection on AWS

With cloud-based architectures, it is important to protect your resources and ensure that enterprise security requirements are met. Network and application security both share a common goal of protecting your resources against cybersecurity threats.

Network security on the cloud is different from that of on-premises, where the cloud provider is generally responsible for securing the cloud itself, such as physical security of the data centers, maintenance, and updates to hardware. There is a shared responsibility model to protect the cloud network and define a security baseline. Using Amazon VPC, security groups, **Network Access Control Lists (NACLs)**, AWS **Web Application Firewall (WAF)**, and AWS Network Firewall all offer points of network protection for your AWS workloads.

Application security on the cloud is an evolving approach where the AppSec responsibility is generally taken on by the developers, called DevSecOps. Security issues at the application level can include threats such as unauthorized access to application functionality or data, exposed application services due to misconfigurations, and hijacking of user accounts because of poor encryption and identity management.

One of the most common attacks is the **Distributed Denial of Service (DDoS)**. There are three types of DDoS attacks:

- **Application-layer attacks**: These refer to well-formed but malicious requests in the form of HTTP GETs requests and DNS queries. These are designed to consume excessive application resources, thereby depleting memory and blocking other legitimate requests from being served. An example of this attack is opening up multiple HTTP connections and reading responses for a duration of seconds or minutes.

- **State-exhaustion attacks**: These refer to the abuse of stateful protocols, which put stress on firewalls or load balancers by consuming an excessive number of per-connection resources.

- **Volumetric attacks**: These block networks by flooding them with voluminous traffic.

AWS services for host-level protection

Information about Amazon VPC security groups can be found at this link: `https://docs.aws.amazon.com/vpc/latest/userguide/security-groups.html`

You can control incoming traffic using security groups at no additional charge by configuring the security group rules. Think of this as a firewall or guardrail to grant permission for a particular type of traffic. The default security group grants permissions to allow all communication and the rules will enable desired communication between the instances only.

AWS Firewall Manager

Firewall Manager is a security management service to centrally configure and manage firewall rules across your accounts in AWS Organizations. For the SecOps teams that do the heavy lifting of getting a distributed but centralized control, Firewall Manager can make it easier to centrally deploy baseline security group rules and protect their VPCs.

When it comes to working with security groups, here are the best practices that AWS recommends:

- Remove unused or unattached security groups to avoid any misconfiguration and create confusion.

- Limit modification to authorized roles only to avoid any unauthorized changes.

- Monitor the security groups and their creation and deletion. This best practice goes hand in hand with the previous two recommendations.

- Limit the ingress or inbound port ranges that need to be accessible. This helps in avoiding exposure to any vulnerabilities or unintended access to services.

All the previously mentioned guardrails can be an overhead for the application owners to incorporate for large-scale AWS environments. This becomes especially challenging when they have to do this every time new applications get deployed or managed centrally on a regular basis.

With *Firewall Manager*, centrally configuring and managing firewall rules becomes easier. Bringing your accounts and applications into compliance by enforcing a set of baseline security safeguards your entire infrastructure.

Here are the core features of Firewall Manager:

- Central deployment for security administrators to enforce baseline and mandated firewall rules across an organization. Firewall Manager helps with reporting any non-compliant issues that are missing required protections.

- Seamless integration to apply AWS WAF rules on ALBs, API gateways, and Amazon CloudFront accounts. You can use Firewall Manager to create a common primary security group across your EC2 instances in your VPC. You can associate your VPCs with Route 53 Resolver DNS Firewall rules as well as automatically create new resources and be notified about them.

- Cross-account protection policies to group resources across accounts are possible with Firewall Manager and AWS Organizations integration. Once you build the protection policies, you can define a group of resources and associate them with your policy. You can specify the scope of the policy to cover the set of AWS accounts or all accounts under AWS Organizations. Firewall Manager can deploy these protections on the resources in the accounts based on the scope of the policy.

- Multi-account resources can be grouped by tags, resource type, or account. Then, security administrators can create policies for these resources with that specified group or across accounts in the organization.

- AWS Marketplace support allows you to centrally deploy and monitor Marketplace-subscribed third-party cloud firewalls across all the VPCs in your organization. This is useful when you have to deploy and manage both AWS-native firewalls and AWS Marketplace-subscribed third-party firewalls.

AWS Network Firewall

Many organizations are facing increased threats of cyberattacks such as malware, botnets, and DDoS attacks. Software-based firewalls are becoming more popular on the cloud. AWS Network Firewall is a managed firewall service that protects all the resources across your VPCs. Network Firewall integrates well with the AWS ecosystem and offers several features, such as stateful and stateless inspection, intrusion prevention, and web traffic filtering capabilities.

When it comes to differentiating Network Firewall from AWS WAF, the latter can handle all types of traffic and not just network traffic. Network Firewall's abilities, such as deep packet inspection, application protocol detection, domain name filtering, and providing an intrusion prevention system, make it favorable for companies that are looking to fulfill network protection and access prevention requirements without upfront investment and overhead costs of managing third-party applications.

AWS Network Firewall sits in front of your VPC, where it will inspect and control VCP-to-VPC traffic. It logically separates networks that host sensitive applications and performs traffic filtering to prevent data loss. Network Firewall secures Direct Connect and VPN traffic from client devices and secures traffic coming from AWS Transit Gateway. Let us look at some additional capabilities of Network Firewall.

Information about AWS Network Firewall can be found at this link: `https://aws.amazon.com/network-firewall/`

AWS Network ACLs (NACLs)

When deploying resources on AWS, one of the most important tools available in the AWS security offerings is the NACL. An NACL is a firewall for your VPC to control traffic in and out of one or more subnets. This is not only an additional layer of security to the DiD but also assists in incident response. NACLs and security groups are often looked at as similar concepts; however, there are differences that can help understand both of these concepts:

	NACLs	Security groups
Purpose	Protect subnets	Protect instances
Stateful versus stateless	Stateless: must specify both ingress and egress	Stateful: return traffic is always allowed, regardless of the rule
Operating level	Process rules in number order	Evaluates all rules
Allow versus deny	Both allow and deny	Allow rules only
Application	Automatically applies	The application must be specified

Table 10.1 – NACLs versus security groups

When it comes to NACL usage, they are most effective for filtering external traffic to internal subnets and more useful when used to insert traffic controls between subnets. Here are a few best practices when using NACLs:

- Use NACLs as a secondary line of defense. Be mindful of default NACLs and use them cautiously, especially if you are running a production server.

- Allow all outbound traffic and deny all inbound traffic by default. Keep a continual check on NACLs that allow all inbound traffic.

- Keep NACLs simple and only use them to deny traffic wherever possible.

- Build your security group rules into your NACLs to reduce the attack surface of your applications.

- Keep a continual check on unrestricted outbound traffic on NACLs and limit access to the required ports or port ranges.

- Log your NACL events using the Amazon CloudWatch logs group to audit your rules regularly and get rid of unused ones.

- Don't overlap multiple subnets by ensuring that the **Classless Inter-Domain Routing** (**CIDR**) blocks for each subnet are different.

There are NACL limitations in terms of default rule limits for inbound and outbound. For large-scale multi-account environments, it is beneficial to leverage a combination of egress firewalls to control outbound traffic and segregate network traffic among VPCs.

AWS Shield

AWS Shield is a managed service to protect web applications against DDoS attacks. This service can be used in conjunction with ELB, Amazon CloudFront, and Amazon Route 53 to protect from DDoS attacks. *AWS Shield Standard* and *AWS Shield Advanced* are the two available offerings, and as the name implies, Shield Advanced comes with a lot more protection than the Standard version at an additional cost.

The Shield Standard provides protection against some of the more common layer 3 (the network layer) and layer 4 (the transport layer) DDoS attacks. Shield Advanced provides additional DDoS mitigation capabilities with intelligent threat detection. Shield Advanced protects your applications at layer 7 and other application layer attacks such as **SQLi**, **cross-site scripting** (**XSS**), **remote file inclusion** (**RFI**), and other threats identified in the **OWASP** Top 10 publication.

Here are some of the features of AWS Shield:

- **Traffic monitoring**: Shield inspects the incoming traffic for you and applies a combination of algorithms and techniques to detect malicious activity.

- **DDoS mitigation**: AWS Shield guarantees that over 99% of infrastructure layer attacks are automatically mitigated in less than a second.

- **Global threat dashboard**: All the information about DDoS attacks on the AWS network is captured on the global threat dashboard.

- **Easy to use and cost-effective**: Shield integrates seamlessly within the AWS ecosystem, which makes it easy to use, and no additional charges are applied for enabling the Standard version of protection. AWS Shield Standard is automatically enabled for all AWS customers at no additional cost.

- **Advanced real-time metrics and reports**: Shield Advanced provides real-time reports with AWS CloudWatch metrics and attack diagnostics.

- **Cost protection for scaling**: Shield Advanced protects you from bill spikes when your accounts are compromised by a DDoS attack that results in scaling your infrastructure.

- **24x7 access to AWS DDoS Response Team (DRT)**: DRT is available for custom mitigation techniques during attacks.

In many cases, AWS Shield Standard is sufficient to meet small and medium-sized businesses' security requirements. If your organization is running complex IT systems or is a likely target of large DDoS attacks where you need to employ custom logic to protect against layer 4 and layer 7 attacks and avoid additional overhead to mitigate sophisticated DDoS attacks, then you can try Shield Advanced service to get near real-time visibility into the attacks.

AWS WAF

AWS WAF helps you protect against common web attacks that can impact your application availability or consume excessive resources. The main difference between WAF and Shield is that Shield protects the infrastructure layers of the **Open Systems Interconnection** (**OSI**) model while WAF protects the application layer. You can create security rules that control bot traffic and block common attack patterns such as SQLi or XSS.

Let us look at the core features of WAF:

- **Web traffic filtering**: WAF provides capabilities to filter web traffic with the help of criteria such as IP addresses, HTTP headers and body, or custom **Uniform Resource Identifiers** (**URIs**). This helps in blocking web layer attacks. You can create a centralized set of rules that you can deploy and reuse easily.

- **Bot Control**: WAF Bot Control is a managed rule group that provides visibility and control over bot traffic from consuming an excessive number of resources and causing downtime. You can easily block or rate-limit bots, scrapers, and scanners, and monitor such bots closely.

- **Fraud prevention**: You can prevent account takeovers by monitoring your application's login page for unauthorized access to user accounts using compromised credentials. Anomalous login activities such as brute-force login attempts and credential stuffing attacks can be avoided, and you can protect your application against automated login attempts by bots.

- **Real-time visibility**: WAF provides real-time metrics and captures raw requests that include details about IP addresses, geo-locations, URIs, and user agents. WAF integrates with Amazon CloudWatch, which makes it easy to set up custom alarms and define thresholds.

- **Integration with AWS Firewall Manager**: You can use WAF centrally across your AWS accounts through AWS Firewall Manager if you are already using it. Automatic audits and security alerts when there is a policy violation for you to remediate immediately are possible through this combination of services.

Data protection using AWS

Employing privacy controls and foundational best practices to strengthen your security posture for your workloads is crucial on the cloud. AWS offers technical, operational, and contractual measures to protect your data. This section will cover some of the foundational best practices and privacy controls to apply to your workloads.

Data classification

Classifying data is a way to categorize your data based on its criticality and sensitivity. By classifying the data, you are able to determine the appropriate protection controls. The process of data classification involves the following steps:

1. **Identifying the data**: This step lets you understand the type of data your workload is processing. Information such as business processes that are involved, the data owner, and any legal and compliance requirements where it's stored will help you determine the resulting controls that need to be implemented. For more information on the detailed process involved in data classification, check out the AWS whitepaper on data classification – `https://docs.aws.amazon.com/whitepapers/latest/data-classification/data-classification.html`.

   Ensuring S3 buckets are not publicly accessible and configuring the least privilege access are essential steps toward better AWS S3 security. In addition, you should ensure they can only be accessed by those who need the data.

   Data leaks from S3 buckets often occur because a bucket containing sensitive data is configured to allow public access.

   Amazon Macie helps to recognize sensitive data such as **Personally Identifiable Information** (**PII**) or intellectual property. Macie uses machine learning and pattern matching to analyze potential data security risks and enables automated protection against those risks. Macie's automated data discovery allows you to gain visibility into your data residing on Amazon S3.

   With Macie, you are equipped with tools for scanning for the presence of sensitive data, such as names, addresses, and credit card numbers, and continually monitoring for security controls, such as encryption and access policies.

2. **Defining the data protection controls**: Protecting data according to its classification level is the next step that your security team can implement.

    AWS service controls such as IAM policies, AWS Organizations **SCPs**, **AWS Key Management Service** (**AWS KMS**), and AWS CloudHSM can help you implement policies for data classification and protection with encryption.

    AWS KMS allows you to centrally manage cryptographic keys across your applications. You can encrypt the data and perform signing operations with KMS. Having centralized control over the life cycle of your keys enables you to manage them seamlessly.

    AWS CloudHSM helps you meet the contractual and regulatory compliance requirements for securing your data. Accessing your keys on **Federal Information Processing Standards** (**FIPS**)-validated hardware and single-tenant **Hardware Security Module** (**HSM**) instances running in your own VPC is made possible through this service.

- **Defining data life cycle management**: It is key to have a life cycle strategy for your data based on its classification type. The strategy should include aspects such as the duration for retaining and destroying the data, data transformation, and data sharing. Also, consider having multiple levels of access for implementing a secure and usable approach for each level of data.

Always implement a DiD approach and minimize human access to data and mechanisms for data transformation purposes.

Automating the identification and classification of data can reduce the risk of human exposure and errors. Using automation with services such as Amazon Macie, which uses machine learning to automatically discover, classify, and protect sensitive data, can help you with not doing the heavy lifting that is involved in setting up the automation tools.

Protecting data at rest

Data that is persisted in non-volatile storage for a particular duration is data at rest. Block storage, object storage, databases, archives, and IoT devices are all examples of data at rest. Protecting such data will minimize the risk of unauthorized access. Two mechanisms of protecting data at rest are *tokenization* and *encryption*.

Tokenization is the process of defining a token to represent data. Encryption is a way of transforming the data to make it unreadable without a secret key that is used to decrypt the content back to plain text.

AWS services such as AWS KMS can be used to implement a secure key to manage data.

Protecting data in transit

Data that is sent from one system to another via a network is data in transit. Your security administrators can protect the confidentiality and integrity of your data by providing the required level of protection for your data in transit. Securing data between VPCs or on-premises locations can be achieved with AWS PrivateLink. PrivateLink allows you to create a secure and private network connection between VPC or on-premises connectivity to services hosted in AWS. With PrivateLink, you can access services across AWS accounts with overlapping IP CIDRs. PrivateLink also works with third-party solutions as well to create a centralized global network.

Summary

In this chapter, we learned about the security controls that AWS provides and recommendations to protect your information, identities, and applications on the cloud. We also reviewed how improving your core security and compliance requirements using AWS services allows you to shift your focus to innovating your business. We also covered how we can give our security teams a single pane of glass to automate and reduce risk with the ecosystem that AWS offers.

In the next chapter, we will explore cloud networking and how it has evolved over the years. We will cover networking fundamentals on the cloud and the various cloud offerings to design your applications for high availability and reliability.

11

Transforming Networking on the Cloud Using AWS

Cloud computing has undoubtedly become a transformative force in the IT world. Resource sharing is the primary focus of cloud computing and networking is one of the core components to enable cloud computing. With the advent of cloud services, networking can empower data centers to unilaterally provision computing capabilities such as storage, OS, servers, and many more. Contrary to traditional data centers, virtualization and cloud computing touch each and every component that can exist anywhere on any server. That makes networking a strategic element for the delivery of cloud-based services to connect the systems and provision and scale the resources to meet end user requirements.

Networking on the cloud needs to be robust as the computational density increases in the cloud environment. Thus, it is essential for organizations to transition to the cloud effectively and architect their networking with the appropriate controls, routing, application performance, and security technologies. In this chapter, we're going to cover the following main topics:

- Introduction to networking on the cloud
- New generation connectivity needs
- Strategies to modernize the network – why and how?
- Networking on the cloud using AWS

Introduction to networking on the cloud

Traditional corporate networks that host their applications using on-premises servers, routers, and switches can no longer keep up with the growing demands of their businesses and compete with the scalability, reliability, or security that the cloud has to offer.

When it comes to a cloud network, there are a few characteristics that differentiate it from on-premises:

- **Isolation** – Cloud networks typically share the same underlying physical infrastructure. Network isolation is a crucial aspect of ensuring security. Cloud providers strive to maintain network architecture that leverages standardization and the safe sharing of pooled resources. A proper network design allows for effective resource management and network security. User traffic is isolated to provide an initial layer of security and higher bandwidth to support tiered networks.

- **Connectivity** – Cloud connectivity is the internet connection between a network and the cloud. Hosting on the cloud is broadly classified into hybrid, public, and private, as we discussed in *Chapter 2*. Each cloud provider uses their models to configure networks or connections across your resources on the cloud. Cloud connectivity includes using the internet or a dedicated private connection such as AWS Direct Connect. There are three main types of connectivity:

  - **Network-to-VPC**: This option connects remote networks with your VPC environment. This is typically used for integrating cloud resources with your existing on-site services.

  - **Software remote access-to-VPC**: This option enables network connectivity between two or more hybrid connection points. This is typically preferred by small companies with a low number of remote networks that need to provide remote access solutions to their employees.

  - **VPC-to-VPC**: This design pattern is to integrate multiple VPCs into a large virtual network. This is ideal for integrating aspects such as billing, security, and so on across multiple VPCs.

  Each cloud provider has its own design and a broad mix of network models to provide you with the advantages of low cost, increased performance, and faster turn-up.

- **Traffic filtering** – Blocking threats is not an easy task for any company. It involves time-intensive efforts, such as the daily analysis of monitoring inbound and outbound traffic to identify bad actors at play. Cloud-based traffic filtering can protect against malicious sites, and allow or deny access to sites based on your lists. Since the connections involve filtering, this gives relief to the engineers, who get increased control over the communications that occur on the network.

In the next few sections, we will explore the constructs of network infrastructure on AWS and how your business can rely on cloud networking to connect users and data centers and run mission-critical applications.

New generation connectivity needs

The dynamics of cloud computing are changing. The strain on networking is increasing as we put more emphasis on virtualization and cloud computing technologies. With enterprise organizations actively moving toward cloud services, the objectives of the organization's network infrastructure and network strategy are rapidly changing. Understanding how cloud computing impacts the network and networking team is crucial across midmarket and enterprise organizations. Let us look at the key networking aspects of cloud computing and the future needs to succeed.

Highest network availability

Next-generation networks operate 24x7, so downtime is not an option. Having a highly available system ensures the delivery of a positive user experience. Application availability is highly dependent on network agility. Having a long-term and fully automated agile framework for the network is crucial to achieving network agility. The key recommendations by industry experts include removing unnecessary complexity and simplifying operations. It is important to build networks that have self-healing capabilities and allow for transparent in-service software updates. Additionally, embracing automation to build a dynamic and responsive network infrastructure is a key aspect of maintaining the highest availability levels for any enterprise's mission-critical workloads.

AWS uses Regions, which are divided into separate Availability Zones and have their own power and network connectivity. The recommended approach is to run your enterprise applications in more than one Availability Zone model to ensure high availability. This helps in isolating from failures that are catastrophic in nature.

Broadest global coverage

Having a modular and scalable network framework will empower your businesses to optimize and address any short-term concerns effectively. This helps in establishing a robust foundation so you can deliver on the evolving needs of consumers and your business. Network elasticity is one of the primary attributes of a flexible modern network. The ability to upgrade or downgrade capacity effortlessly can empower your business to respond quickly to changing business needs.

AWS offers 84 Availability Zones across 26 Regions that offer low latency and high throughput and are highly redundant (`https://aws.amazon.com/about-aws/global-infrastructure/regions_az/`). Enterprises can leverage more than 310 points of presence, which provides global coverage for your end users. Instances can offer up to 100 Gbps of network bandwidth.

Guaranteed high performance

Strong network performance with low latency and high availability is key for improving the speed of IT service delivery. Studies suggest that network performance is one of the top reasons for companies to move to the cloud. Cloud networks are designed to achieve microsecond latencies. This implies improved application performance and better observability in providing predictable performance. Unlike the on-premises network monolith stacks, you get seamless consistency across the network fabric. AWS offers a broad range of services when it comes to **high-performance computing** (**HPC**), which can be a great start to finding the best-performance components at an optimal cost. If you have workloads that run large and complex simulations that require thousands of CPUs and GPUs, you can leverage AWS's HPC services such as **Elastic Fabric Adapter** (**EFA**) and AWS ParallelCluster.

EFA is a network interface for Amazon EC2 instances and is built on a platform called AWS Nitro System, which offers a rich collection of building blocks such as Nitro Cards, Nitro Security Chip, Nitro Hypervisor, Nitro Enclaves, and Nitro TPM. AWS ParallelCluster is an open source cluster management tool to deploy and manage **HPC** clusters on AWS. These offerings make it easy to run simulations for a wide variety of use cases, such as genomics, computational chemistry, financial risk modeling, machine learning, and deep learning models.

You can run your high-performance applications by choosing from a variety of compute instance types, such as third-generation Intel Xeon processors, Graviton3 (the latest in the AWS Graviton processor family), and NVIDIA GPU-based instances, which are ideal for compute-intensive workloads.

Most secure

The need to secure is a top priority for your teams while connecting from any location around the globe. Having a holistic security strategy that eliminates complexity is important. Many companies rely on security solutions such as VPNs, secure web gateways, and network firewalls but can also create choke points that can hurt your application's performance and employee productivity.

AWS offers network and application protection services to enforce a fine-grained security policy at every checkpoint across your organization. By using these services, you get complete in-line control of your traffic, which can help you protect against unauthorized access, potential vulnerabilities, and performance degradation.

Network infrastructure modernization strategies

As businesses grow in this digital transformation era, it is seemingly becoming important to modernize your network infrastructure and meet the changing demands of the future. In this section, you will take a closer look at network infrastructure modernization, its benefits, and some proven ways to successfully modernize your network.

What is network infrastructure modernization?

In the past, applications often took several years to be fully developed and were dependent on physical servers that were hosted in corporate data centers. Modernizing network infrastructure at its core entails replacing your existing legacy network fabric with modern and adaptable solutions. Solutions typically include implementing physical and digital policies, processes, and technologies to enhance your current networking capabilities to meet the current and future demands of your business.

The cloud era propelled many enterprises to treat network modernization as an essential step to achieve efficient operating practices. These practises extend from the core, made up of data center facilities and clouds, to remote sites worldwide. As per the **2022 Environmental, Social, and Governance report** (https://www.redcross.org/content/dam/redcross/about-us/publications/2022-publications/Environmental_Social_and_

`Governance_Report_2022.pdf`), more than 60% of organizations said that they want to modernize their network to become operationally efficient and deliver a better customer experience. With the evolution of microservice architecture and container platforms, applications can be broken into smaller pieces and hosted in cloud-based agile and application-centric IT environments using DevOps methodologies. While the microservice-based architecture enables development teams to iterate updates and newer functionality at a faster pace, DevOps practices can ensure optimized application delivery in production environments. With these changing trends, businesses witness benefits, and the role of application developers and **site reliability engineers** (**SREs**) is becoming more significant in driving positive customer experiences. This transformation can be supported by a highly distributed and dynamic virtualized environment that can rapidly spin up services wherever needed to accelerate the phases of the application life cycle.

The role of the modern network is becoming ever-important to businesses to contend with the highly distributed environment and have an efficient application-centric modern network. With increasing IT complexity and the evolving landscape of microservice architectures, businesses are opting for software-driven enterprise networking environments. **Software-defined networking** (**SDN**) has emerged as an architectural approach.

Advantages of modernizing your network infrastructure

A successful modern network environment can support businesses with the following capabilities:

- **Operational efficiency**: With a modern network infrastructure, organizations can better manage their distributed applications through easy-to-use environments and simplify complex tasks that are often manual. Employees no longer are restricted by the limitations of legacy network infrastructure and can have more time to learn new technologies and improve their skill set. Automated patching and updates remove potential risks that are often caused due to human error, thereby improving agility and productivity.

- **Faster time to market**: Modern networks enhance agility in terms of provisioning network connectivity in a dynamic manner. This enables applications to be built faster and businesses can achieve competitive differentiation.

- **High availability**: Modern network infrastructure is designed with self-healing and self-scaling capabilities that can help you achieve greater application availability and zero downtime. Features such as automated life cycle management can enable you to apply patches and upgrades without having to take your services offline.

- **Seamless connectivity**: The public cloud service supports a true hybrid cloud or multi-cloud approach. This helps to promote seamless and optimized connectivity for employees who can work from anywhere. Employees are not restricted by the legacy network infrastructure and can be more productive by learning more skills to improve the business and focus on strategic initiatives.

- **Security enhancement**: Security is an integral part of a highly distributed modern network to deliver a truly digital experience. Network modernization helps ensure your business meets the ever-changing complex compliance requirements needed to adhere to industry standards, such as the **General Data Protection Regulation (GDPR)** and **Health Insurance Portability and Accountability Act (HIPAA)**. Data encryption from the edge to the cloud to the data center enforces governance and compliance to better protect the end users and the applications they access.

- **Cost reduction**: High availability decreases the revenue loss from any unexpected downtime and performance degradations, which typically is the case for legacy systems. This also reduces maintenance costs and business disruptions, thereby creating an opportunity to eliminate unnecessary spending while meeting the requirements of your business as it scales.

- **Better end user experience**: Modern networks enable end users to experience a seamless application experience. With the digital transformation initiatives that enterprises are adopting, ensuring a consistently positive experience is a critical factor to determine success or failure. Additionally, employees are empowered to work remotely, with seamless, secure, and optimized connectivity, which essentially creates a network to drive greater job satisfaction.

In the next section, we will learn how organizations can evolve into modern networks and ensure their networks can keep pace with their modern applications and services.

Strategies to modernize network infrastructure

Businesses are focusing on their networks as they adapt to the cloud to make sure that their enterprise backbone is meeting their digital transformation demands. Here are the key steps to accelerate your journey toward network modernization.

Understand your existing infrastructure and identify the gaps

As network administrators, evaluate your existing **local area network (LAN)**, **wide area network (WAN)**, internet, and cloud components thoroughly. As shown in the following diagram, at home, your **LAN** connects devices over tens of meters. A **WAN** operates over a much larger area as it connects LANs to enable data exchange.

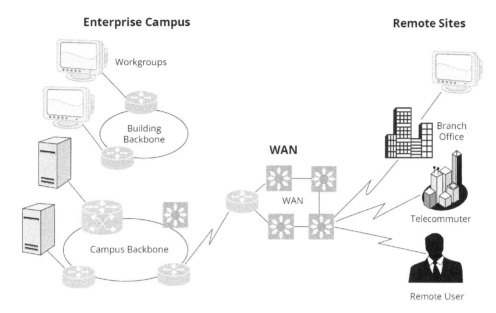

Figure 11.1 – LAN, WAN, and the internet

Collaborate with the application teams and perform the analysis at the application level first, and then traverse through the existing network and data. Document the data and network flows for your business-critical applications. This will help you identify the performance gaps and opportunities for cost improvements. Highlight all the areas where new hardware, platforms, or even managed services can boost performance and drive cost efficiencies. This will enable you to make a strong case for the modernization of your network infrastructure with your leadership.

Let's take a scenario for a certain application where employees are increasingly working from home due to the COVID pandemic. Evaluate all the traffic from the corporate LAN network and identify security gaps. Understanding how these gaps can be addressed by leveraging modern secure access service edge tools will enable you to make a better case for extending your existing IT security policies, as well as better protect sensitive data.

Upgrading network technologies

Many businesses have dated network equipment that cannot scale or address future requirements. Start evaluating the equipment along with the 802.11 wireless standards and wireless technologies, if they don't offer great throughput or connectivity. Also, examining the access switches and replacing them with multi-gigabit Ethernet access with modern capabilities to capitalize on cabling infrastructure is a good approach. Examining your cabling infrastructure to support the growing bandwidth requirements such as 10 gigabit/second speeds and the newest OM5 wideband multimode fiber optics is also important.

As your system applications, data, and services grow, the traffic is going to put a strain on the networking routes. Moving to the next generation of wired networking, such as 10/25 Gbps server networking, will open up faster network paths to handle the data and application needs. On the wireless side, Wi-Fi access points are becoming more and more reachable, and you can leverage and provide high-speed wired networking.

Finally, consider redesigning your network architecture to make sure it is open and flexible.

Networking on the cloud

It is always a daunting task to understand what is running on your current infrastructure. Knowing the list of workloads that are at risk and the state of their performance, resiliency, and cost of operations will help you determine whether your applications can be supported in a cloud environment and whether they will be easily migrated. Before you begin your migration processes, have your teams capture the size of the VM that they will need to run each application, CPU, and other requirements that the physical servers have. This is also a perfect time for you to consider re-architecture approaches, such as microservices or containers.

Here are some scenarios for you to consider:

- Scenario 1 – Applications for internet-facing purposes:

  - You will need to consider network rules to enable the server hosting the application so that it can connect with the internet.

  - You will need specific ports to connect to the database and still be protected from the internet.

- Scenario 2 – Applications with hardcoded IP addresses:

  - You will need to redesign the application so that hardcoded IP address dependencies are removed and hostnames are considered instead.

  - Or you will need to design the network to support Virtual Network (VNet) addresses if running on AWS, which maps back to the specific IP addresses.

- Scenario 3 – Databases that need virtualized platforms:

  - If you need to connect to a persistent layer, you will want to consider the amount of data and I/O that the database needs to have.

  - This is the right time for you to think about traditional databases versus NoSQL databases.

Additionally, if you are moving from one physical environment to a virtual one, consider building a new image that is virtual or cloud-native and is up to date on the OS.

Invest in network automation

Automation simplifies the burden of manual tasks that are often redundant and tedious. This not only helps in freeing up the IT staff's time but also reduces the need for human intervention. Automation is considered widely across all modern IT environments and is a critical element for digital transformation. For many organizations, network automation is a critical yet missing link to achieving an automated, self-operating, and predictive stack. As networks are becoming more and more complex, there is a need to have a single pane of glass for multiple platforms catering to both on-premises and cloud purposes.

Having an intelligent network automation that is built intentionally to react quicker when unplanned events occur, such as security threats or vulnerabilities, ensures business continuity. Consider the differences in modern infrastructures, such as API integrations. Additionally, ensuring compliance and preventing configuration drift becomes manageable with automation. Automation is also useful for tracking network configuration changes to analyze or identify potential threats, especially across a network that consists of thousands of devices.

In the next section, you will learn about cloud networking: the broadest set of networking offerings that can be leveraged while running your workloads on the cloud with AWS.

Networking on AWS

The networking services offered by AWS constitute the backbone of the workloads that you run on the AWS global cloud network. These services offer the highest level of security, reliability, performance, cost-effectiveness, scalability, and global coverage for your applications. In this section, I will cover the key constructs and break them down from a networking perspective.

Network foundations

AWS's fundamental constructs form the foundation that creates a sustainable AWS network to support an enterprise-level organization.

Amazon Virtual Private Cloud - For any organization, a strong network setup is a foundational block. **Amazon Virtual Private Cloud** (**Amazon VPC**) is a foundational service that is critical to run your workloads in the AWS cloud. With Amazon VPC, you can quickly spin up an isolated virtual network environment in your AWS accounts. Think of it as a traditional TCP/IP network that can scale as needed. The components that you deal with on-premises such as routers, switches, VLANs, and so on do not exist in a VPC, as they are all abstracted and re-engineered into the AWS data centers.

You configure IP addresses, subnets, routing, and security for each VPC and, once created, the VPC lives in a specific AWS Region that is geographically closer to your customers, thereby reducing any latencies. VPCs are isolated from each other, which means that you can create private subnets with either IPv4 or IPv6 address blocks. A **subnet** contains a subset of the VPC Classless Inter-Domain Routing (CIDR) block and isolates its traffic from the remaining VPC subnet traffic.

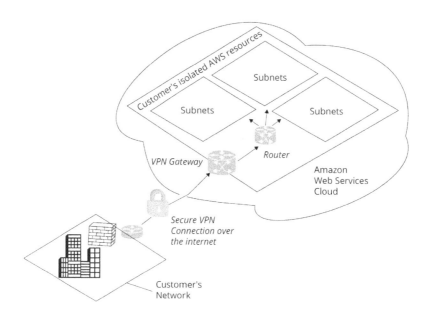

Figure 11.2 – Amazon VPC

Route tables determine how the incoming traffic should be directed inside your VPC and subnets. Internet connectivity can be configured using the internet gateway, NAT instances, or a NAT gateway.

Traffic monitoring within VPC is used to monitor the network traffic of the applications. The users can have direct access to the network packets that travel through a VPC. This traffic data (that is, **VPC Flow Logs**) can be further sent to the security and network analysis tools for inspecting and identifying potential threats and troubleshooting. Alternatively, you can use **VPC Reachability Analyzer** to do connectivity testing, troubleshoot connectivity issues, verify that the network configuration matches the intended connectivity, and automate the verification of the connectivity intent.

Secure connectivity typically requires you to perform complex configurations and set up additional security controls. In order to meet your organization's security requirements, AmazonVPC provides security groups and **Network Access Control Lists** (**NACLs**) to perform inbound and outbound traffic filtering. AWS Firewall Manager is a managed service to deploy additional network protections for your VPCs. AWS Firewall Manager provides flexible rules to give you fine-grained control over network traffic, such as blocking outbound requests and protecting you from network threats.

To cater to scenarios that require connecting two VPCs privately to get more flexibility and connect without the need for the internet, **VPC peering** can be used. A VPC peering connection enables connection between two VPCs to route traffic between them via private IPv4 or IPv6 addresses. This provides the advantage of low cost since you only need to pay for the data transfer and at no bandwidth limit.

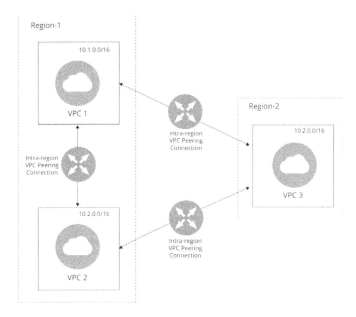

Figure 11.3 – AWS VPC peering

As you can see in the preceding diagram, you can create a VPC peering as a one-to-one connection between your own VPCs, with a VPC in another AWS account, or with a VPC in a different AWS Region. You get the ability to create multiple VPC peering connections for each VPC to ensure there is no bandwidth bottleneck or a single point of failure.

Scalable networking using **AWS Transit Gateway** when you are expanding globally and creating hundreds or thousands of VPCs simplifies your entire network and routes the traffic to and from each VPC with monitoring. Transit Gateway acts as a central hub to simplify your network and puts an end to complex peering relationships. The network manager feature allows you to easily monitor your VPCs and edge connections from a single pane of glass. You get a unique view over your entire network and it even helps you to connect with **Software-Defined Wide Area Network (SD-WAN)** devices.

What can you do with a VPC? Here are some use cases:

- **Launch a simple website or a blog**: You can host a public-facing website, such as a simple tiered web application or a blog, by creating a public subnet and securing it using firewalls, known as security groups, which allow inbound traffic, either HTTP or HTTPS.

- **Host multi-tier web applications**: A multi-tier web application requires strict access controls and restricted communication between the multiple layers of the application, such as the web server layer, application layer, and database layer. Amazon VPC has all the functionalities to facilitate access controls such that the inbound and outbound access for internet traffic is configured optimally.

- **Create hybrid connections**: Many businesses have requirements for connecting their multiple branch locations through interconnected networks. Amazon VPC provides an abstraction layer to facilitate building hybrid networking solutions.

Application networking

Application networking using AWS services helps you to improve the overall application network architecture. You can monitor each component of your application using AWS CloudWatch's metrics, logging, and request tracing. Streamlined network monitoring, support for global applications, and improving your application availability and performance while building robust security into your applications are all possible with the services offered for application networking.

AWS **Elastic Load Balancing (ELB)** is a key architectural component for AWS-powered applications that are used to distribute incoming application traffic across multiple compute resources. As a result, your application's availability and fault tolerance will increase in response to the incoming application traffic.

In order to achieve the highest level of availability, ELB supports health checks for each service independently.

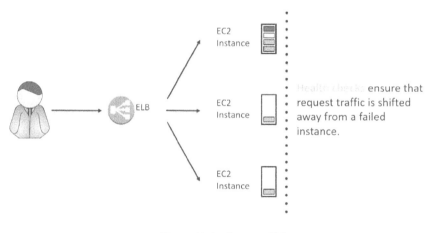

Figure 11.4 – Amazon ELB

There are few types of load balancers:

- **Classic Load Balancer (CLB)** is primarily used to balance services for the EC2 instance network at level 4 of the **Open Systems Interconnection (OSI)** model. Many web applications typically use the TCP/IP protocol at the level 4 transport layer and UDP in some cases. CLB uses the information for incoming requests and routes the traffic to the EC2 instances that are hosting your web applications. CLB makes routing decisions at either the transport layer or the application layer.

- **Network Load Balancer (NLB)** supports TCP, UDP, TCP+UDP, as well as TLS listeners. NLB is capable of handling millions of routing requests between clients and target systems using IP addresses, TCP, or port numbers. NLB offers robust load balancing for volatile requests. With the support of dynamic host port mapping, traffic is distributed to the workloads.

- **Application Load Balancer (ALB)** supports targets with any OS. Any applications using HTTP or HTTPS protocols are supported. ALB can route traffic to modern application architectures such as containers, IP addresses, EC2 servers, Lambda functions, and microservices. ALB supports Lambda functions and containerized applications and with dynamic autoscaling, the independent application services can be monitored independently.

- **Gateway Load Balancer** supports inline virtual appliances where network traffic transparently passes all layer 3 traffic through third-party virtual appliances. Virtual appliances such as firewalls and intrusion, detection, and prevention systems can be deployed at scale. Gateway Load Balancer operates at the third layer of the OSI model and provides private connectivity between virtual appliances and the service provider VPC.

You can select the appropriate load balancer based on your application needs. The choice mainly depends on your infrastructure environment, cost, security, and how your traffic must be handled between the clients and the targets. If you have to balance HTTP requests, go with ALB. For network/transport load balancing, use NLB. And if your application is built within the EC2 classic network, go with CLB. If you need to run third-party virtual appliances, you can choose Gateway Load Balancer.

AWS Global Accelerator improves the overall availability and performance of your applications when you put them in front of your ALB. This service uses the AWS global network to direct end users' traffic to a healthy application endpoint in an AWS Region that is closest to your customer, thereby improving the latency and throughput of your application traffic.

AWS App Mesh provides application-level networking so your services can communicate with each other across different types of AWS compute offerings, such as AWS Fargate, Amazon EC2, Amazon ECS, Amazon EKS, and Kubernetes running on AWS. This service helps you capture logs, metrics, and traces from all your applications, and combined with Amazon CloudWatch and AWS X-Ray, you can quickly identify and isolate issues with any service within your entire application. All the requests can be encrypted between the services, and you can implement custom traffic routing rules to ensure that your service is highly available during deployments. The following diagram shows how App Mesh provides application-level networking capabilities for the services to communicate across multiple layers of compute infrastructure.

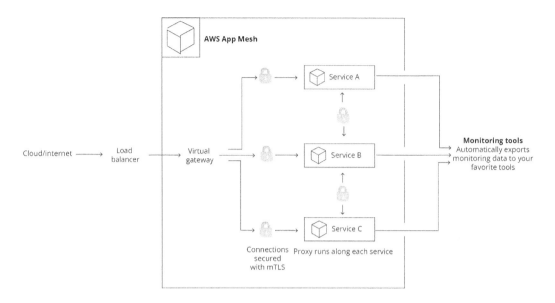

Figure 11.5 – AWS App Mesh

App Mesh uses an **open source Envoy proxy** to manage the incoming and outgoing traffic of the service's containers. The Envoy proxy is a high-performant C++ distributed proxy designed to serve as the communication bus for large microservice "service mesh" architectures. Here are the App Mesh features:

- **Traffic routing** – App Mesh makes it possible for you to dynamically manage traffic routing between services without any application code changes.

- **Mutual TLS (mTLS)** authentication provides service-to-service identity verification for the application components running in and outside service meshes. You can extend the security perimeter to the applications by provisioning certificates from AWS Certificate Manager, AWS Private Certificate Authority, or a customer-managed certificate authority.

    Integrations connect your EKS, ECS, or EC2 instances to App Mesh and export metrics, logs, and traces to configure traffic routes and other controls between microservices that are mesh-enabled. App Mesh can handle the traffic between internal services within an AWS Region and provides a greater degree of control and monitoring for service communications.

    App Mesh is also compatible with various AWS partners and open source tools such as Datadog, Alcide, HashiCorp, Sysdig, SignalFx, Spotinst, Tetrate, Weaveworks, Twistlock, and Aqua.

- **Amazon VPC Lattice** is a new service that is in beta currently. This is an application layer service that provides the capability to simplify service-to-service connectivity, security, and monitoring across EC2 instances, containers, and serverless applications. This solves the main pain point for developer personas as well as admins. The shift to service-oriented architecture created a requirement to keep the blast radius as small as possible. Creating an improved and consistent security posture with reliable authentication is crucial.

The following diagram showcases the workings of VPC Lattice where you can define policies for network access, traffic management, and monitoring to connect compute services such as instances, containers, and serverless applications.

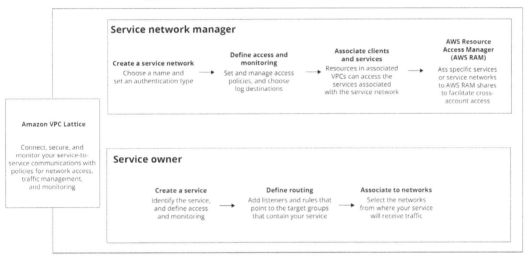

Figure 11.6 – Amazon VPC Lattice

Let us look at some of the core features that make VPC Lattice an ideal choice to operate your network connectivity at the application level.

Service directory is a centralized view of the services that you own or are shared via **AWS Resource Access Manager** (**AWS RAM**).

Service network is a logical application layer network that connects clients and services across different VPCs and accounts. This helps in abstracting network complexity. You can use HTTP/HTTPS and gRPC protocols within a VPC.

Connectivity between VPCs and accounts is to manage network connectivity between VPCs and accounts.

It also helps with traffic management and application layer routing. You get common controls to route traffic based on requests. This helps in weighted routing for blue/green and canary-style deployments.

VPC Lattice also provides context-specific authentication and authorization. Integration with AWS IAM provides the same familiar authentication and authorization experience to the admin teams without having to learn new networking tasks.

In summary, VPC Lattice helps bridge the gap between developers and cloud administrators. You can connect thousands of services across VPCs and accounts without increasing network complexity.

Edge networking

Edge networking allows you to transmit data securely across the world, removing network hops and with improved latency. The traffic is moved off the internet, thereby limiting the exposure to any attacks. The services offered in this space by AWS cater to the following use cases.

Customizable content delivery networking – Amazon CloudFront

Amazon CloudFront is a fast **content delivery network** (**CDN**) that securely delivers data, videos, applications, and APIs to end users globally with low latency and at high transfer speeds. You can speed up the distribution of your static and dynamic web content to the end users. The following diagram shows a simple illustration of how Amazon CloudFront works:

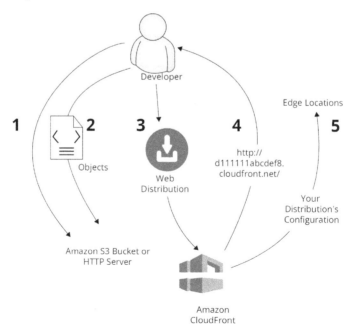

Figure 11.7 – Amazon CloudFront

Amazon CloudFront can be scaled and distributed globally with guaranteed low-latency performance and high availability. As of 2022, per the AWS documentation, Amazon CloudFront has 230+ **points of presence** (**PoPs**) that are interconnected via the AWS backbone. The AWS backbone is a private network built on a global, fully redundant 100 GbE metro fiber network linked via trans-oceanic cables across the Atlantic, Pacific, and Indian oceans, as well as the Mediterranean Sea, Red Sea, and South China Sea.

Amazon CloudFront automatically maps network conditions and intelligently routes your user's traffic to the most performant AWS edge location to serve up cached or dynamic content. CloudFront comes by default with a multi-tiered caching architecture that offers you improved cache width and origin protection.

Global network traffic acceleration – AWS Global Accelerator

As your businesses grow, limited bandwidth on public networks and spikes in network traffic can cause slowed or lost connections. AWS Global Accelerator is designed to avoid a single point of failure and increase operational resiliency. Global Accelerator is a networking service that uses the AWS global network infrastructure to send end user traffic to workloads hosted across multiple AWS Regions and can reduce your dependencies on DNS for failover. Additionally, you can improve your application network performance by accelerating data transfers globally.

Figure 11.8 – AWS Global Accelerator

AWS Global Accelerator has several use cases, such as the following:

- Single region applications, by improving the network routing for local and global user traffic.

- Multi-region applications, where AWS Global Accelerator simplifies the complexity of the traffic routing logic and automatically redirects the traffic to the new connections to a healthy endpoint.

- Gaming, where AWS Global Accelerator improves players' experience by routing the player traffic along the private AWS global network. This also helps in reducing in-game latencies and packet loss.

- **Real-time communication** (**RTC**) – many industries, such as real-time gaming and telecommunication, can benefit from applications including **Voice over IP** (**VoIP**) and video conferencing across a variety of devices.

Reliable network routing and traffic management – Amazon Route 53

Amazon Route 53 is a highly available and scalable cloud **Domain Name System (DNS)** that routes end users to the internet by converting human-readable names into IP addresses. Route 53 combines scalable DNS and health checks designed to give businesses a reliable and cost-effective way to route end users to internet applications.

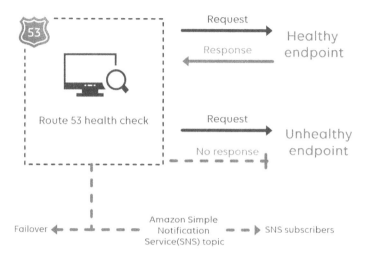

Figure 11.9 – Amazon Route 53

You can create and manage public DNS records, and Route 53 additionally offers health checks to monitor the health and performance of your applications and other resources. Route 53 is designed to provide consistent routing capabilities in a reliable and cost-effective manner and scale, automatically anticipating end user needs.

Hybrid connectivity

Many organizations have realized the benefits of moving their compute infrastructure to the cloud. Although this happening at an accelerated pace, there are few industries that are striving for a consistent experience, from start-ups to large enterprises, to achieve millisecond latencies. Leading government agencies work with large local datasets, and share data with on-premises applications with single-digit millisecond latencies. They are looking to accelerate their digital transformation and operate from a combination of platforms – cloud, on-premises, or at the edge. Running applications on hybrid platforms is complex and challenging when it comes to integrating legacy systems and passing data to and from the cloud.

Through the digital transformation, which often starts with a lift and shift to the cloud for certain applications, here are some of the common hybrid cloud scenarios:

- **Low latency** – Industries that are running financial trading platforms, real-time gaming, electronic design automation, and machine learning inference at the edge would need to maintain minimal latencies to meet their business needs.

- **Local data processing** – Datasets residing locally need to be easily processed and migrated to the cloud due to constraints such as cost, bandwidth, or timing. Organizations seek a consistent hybrid cloud architecture to process data on-premises and move data to the cloud for long-term archival purposes.

- **Data residency** – Industries such as financial services, healthcare, and oil and gas are required to align with security/tax regulations, data sovereignty, and geo-political dynamics.

- **Data center extension** – Organizations that are growing rapidly require their applications to meet use cases such as backup and disaster recovery, which can be achieved via the cloud.

- **Enterprise cloud migration** – Complex enterprise applications need hybrid architectures to operate across both on-premises and the cloud.

AWS provides services with which you can drive the ability to run your applications wherever you need them in the world. Let us take a look at those services and the use cases they serve.

AWS Local Zones

A Local Zone is an extension of an AWS Region with which you can run applications on AWS that require single-digit millisecond latencies to your end users. Local Zones are a powerful construct offering a new type of AWS infrastructure deployment that places compute, storage, and other selected services closer to specific locations.

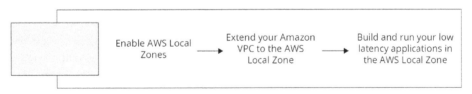

Figure 11.10 – AWS Local Zones

As shown in the preceding diagram, you can easily run applications by bringing AWS infrastructure closer to the end users and business centers. This is a fully managed service supported by AWS, having the same benefits of elasticity, scalability, and security that you get from running on AWS. Let us look at some of the industry use cases that would benefit most from using Local Zones:

- **Media and Entertainment (M&E) content creation**: Many M&E organizations that have live production or video editing manage expensive infrastructure to support millisecond latencies and offer a jitter-free experience. Local Zones can help them run latency-sensitive workloads and achieve as low as 1-2 millisecond latency.

- **Enterprise migration with hybrid architectures**: If you are running complex legacy on-premises applications and would like to migrate to the cloud, it typically involves a lot of steps to migrate a portfolio of applications that have interdependencies to the cloud. With AWS Direct Connect and Local Zones, these enterprises can create a hybrid environment that provides ultra-low latency communication between applications running on Local Zones and on-premises. Migrating incrementally can help in simplifying and enabling ongoing hybrid deployments as a result.

- **Real-time multiplayer gaming**: Gaming companies want to be located physically closer to the gamers' locations to provide real-time experience and avoid any latencies. An ideal gameplay experience requires a latency of 15 milliseconds or less. Organizations that were using on-premises installations can take advantage of Local Zones to run real-time and interactive multiplayer game sessions.

As of 2023, Local Zones have a total of 27 locations present globally, with 17 in the US, 1 in Latin America, 4 across Asia, and 5 in EMEA. Data residency requirements vary depending on the location and jurisdiction. You can configure your data to remain on AWS Local Zones using services such as Amazon EC2, Amazon EBS, FSx, and other local services.

AWS Outposts

Outposts is a fully managed service that extends AWS infrastructure, AWS services, APIs, and tools to virtually any data center, colocation space, or on-premises facility for a consistent hybrid experience. Companies can run what they need locally using the same processor technology that Amazon EC2 instances use. Scalability is possible to the cloud when needed and Outposts helps in removing the heavy lifting required to procure, manage, and upgrade existing on-premises infrastructure. Let us look at some of the industries in which AWS Outposts can enhance the overall experience:

- Healthcare management systems can achieve rapid retrieval of medical information that is stored locally. Analytics and machine learning services can be used on the cloud easily and achieve low-latency processing requirements.

- Telecommunications providers have a requirement to offer reliable network services in multiple locations. Outposts can help orchestrate, update, scale, and manage virtual network functions across the world.

- Media and entertainment companies can use Outposts and seamlessly integrate AWS resources with their on-premises infrastructure and leverage the latest GPU innovations, support live real-time event streaming, and develop cutting-edge livestream gaming experiences.

- Financial services institutions can overcome regulatory challenges and deliver the best digital experiences to customers. They can meet data locality requirements by delivering services from in-country locations.

- Retail industries can deliver next-generation online experiences through Outposts with consistent and reliable operations at every retail location.

- Manufacturing control systems that need to run close to factory floor equipment can run them on AWS Outposts and achieve seamless integration and centralized operations and modernize their applications using AWS infrastructure and services.

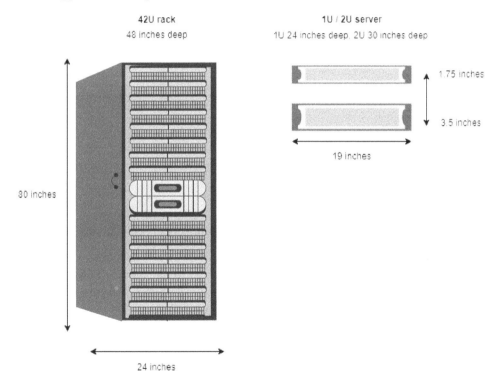

Figure 11.11 – AWS Outposts form factors

As shown in the preceding diagram, AWS Outposts is designed to operate on two different form factors: 42U rack and 1U/2U server.

Outposts can be deployed as full-size racks with compute, storage, and networking or in smaller remote locations. AWS Outposts comprises two broad sets of offerings: AWS Outposts rack and AWS Outposts servers. They are both designed to offer a varied set of power, networking, and service capabilities based on your requirements. The Outposts racks are fully assembled and ready to be rolled into the final position, whereas the Outposts servers will be installed by personnel upon delivery.

VMware Cloud on AWS

Organizations that depend on VMware for vSphere-based workloads can migrate their environment to VMware Cloud on AWS and achieve a simple, faster, and more cost-effective path to a hybrid cloud.

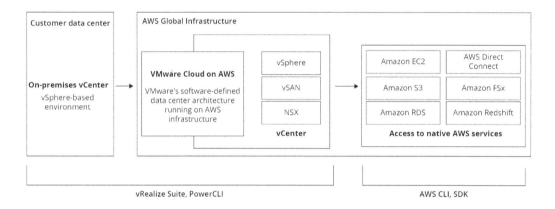

Figure 11.12 – AWS VMware

The preceding diagram showcases a high-level overview of how you can combine compute, network, and storage capabilities.

Your teams can still leverage their VMware skills and experience to run VMware **Software-Defined Data Center (SDDC)** software on bare-metal AWS infrastructure for better performance. Let us look at some of the common ways that you can use VMware Cloud on AWS:

- **App modernization** – Existing enterprise applications can be modernized with minimal disruption. With VMware Cloud on AWS, you can transform to modern frameworks such as Kubernetes or be enriched with PaaS services and DevOps tools for a consistent, upstream-compatible experience that is tested, signed, and supported by VMware.

- **Cloud migration** – vSphere workloads are quite taxing to migrate to the cloud, which typically takes months or years depending on the complexity and interdependencies of the application. With VMware Cloud on AWS, it takes weeks or days and reduces the complexity, cost, and risk of cloud migrations compared to other alternatives that require time and skill sets.

- **Cloud Video Desktop Infrastructure (VDI)** – VDI is a virtualization solution that provides and manages virtual desktops. With VMware Cloud on AWS, you get a cloud platform for virtual desktops and applications to deliver complete VDI infrastructure from the cloud, extending an existing on-premises VDI environment to applications running on AWS.

- **Data center extension** – Organizations often have requirements to add capacity, on-demand scaling of applications, and rapid regional expansion. With VMware Cloud on AWS, you can move your vSphere workloads without changing IP addresses into a VMware-consistent, enterprise-grade environment in the AWS cloud quickly and cost-effectively.

- **Disaster recovery** – With VMware Cloud on AWS, companies can simplify disaster protection and site recovery. You can easily move applications from DevTest to production or burst capacity. Leveraging cloud economics to reduce operational errors is also possible and can help you deliver a highly scalable service.

AWS Wavelength – 5G networks

AWS Wavelength offers enterprises a highly responsive experience on new 5G networks. Data transported by 5G typically needs to hop across multiple networks to reach application servers. AWS Wavelength brings AWS services to the edge of the 5G network with minimal latency when connecting to an application from a mobile device. AWS Wavelength offers ultra-low latency for 5G applications, as well as scale and flexibility, a global 5G network, and a consistent AWS experience. Here are some of the use cases of AWS Wavelength:

- Connected vehicles

- Interactive live video streams

- AR/VR

- Smart factories

- Real-time gaming

- Machine learning-assisted diagnostics for healthcare

In summary, the domains available in Wavelength zones are compute, storage, networking and management, and monitoring.

AWS offers a broad range of services, as we learned in this chapter, to enable hybrid architecture and allow businesses to extend the cloud to anywhere in the world. Choosing the right hybrid architecture and AWS services can empower you to deliver the next generation of applications, services, and capabilities that your business needs and stay competitive.

Summary

In this chapter, we learned how cloud networking is simplified and abstracted. We explored the networking fundamentals that can enable your organization to get started. The cloud is constantly evolving and innovating, and so are the networking best practices. The network teams should keep up to date with the latest practices and evolve to incorporate new features to improve your organization's performance and security.

In the next chapter, we will explore the Well-Architected Framework and the design principles of it. We will also learn about the best practices and guidance on how to approach incident management and streamline SRE practices.

Part 4:
Cloud Economics, Compliance, and Governance

In this part, we will look at cloud operations and how to operate securely and efficiently on the cloud. We will dive deeper into the pillars of the AWS Well-Architected Framework and the best practices you can apply to lower your cost of operations, improve IT infrastructure and staff productivity, decrease operational downtime, and achieve faster time to market. You will learn about operational excellence and the tools that Amazon has developed to operate on the cloud at scale while maintaining governance, security, and compliance for speed, automation, and innovation. We will wrap up by doing a recap of the important topics covered in the prior chapters and discuss emerging trends and technologies to prepare ourselves for the future model of building modern workloads.

This part comprises the following chapters:

- *Chapter 12, Operating on the Cloud with AWS*
- *Chapter 13, Wrapping Up and Looking Ahead*

12

Operating on the Cloud with AWS

Organizations are evolving their teams and processes to keep up with the pace of their cloud adoption and to optimize the cloud transition by creating guidelines, best practices, and alignment. Traditionally, maintaining workloads and operations involved activities such as monitoring network and server performance. With the rise of cloud migrations, IT operations witnessed a shift to activities such as cloud-optimized alerting and monitoring and quickly evolved to **Cloud Operations** (**CloudOps**). CloudOps is a core topic for companies of any size that move to the cloud in a big way. Enterprises have realized the benefits of moving to the cloud, including a lower cost of operations, stronger operational excellence, and a smaller carbon footprint. For many companies, CloudOps is an afterthought, but there is an increasing realization that it is important to have an effective CloudOps strategy for managing the growing complexity of running cloud-based deployments.

Having an action plan is key for various aspects, right from problem management to monitoring, security, and performance. As per the *Bridging the Cloud Transformation Gap* report (`https://info.aptum.com/cloud-impact-study-1#whichcloud`), although 72% of organizations view the cloud as a driver of increased efficiency only 33% of organizations maximized their value while moving to the cloud. Some of the more progressive pioneers realize that CloudOps-related skills development is crucial to implementing operating models.

In this chapter, we're going to cover the following main topics:

- Introduction – getting started with CloudOps
- Building a **cloud center of excellence** (**CCoE**)
- Are you well architected on **Amazon Web Services** (**AWS**)?
- **Cloud Financial Management** (**CFM**) on AWS
- **Site reliability engineering** (**SRE**) on AWS
- Incident management

Let's get started!

Introduction – getting started with CloudOps

CloudOps is a methodology to manage IT operations such as delivery, optimization, and performance of workloads that are running in a cloud environment. The principle of DevOps is extended in CloudOps to optimize cloud resources and procedures to achieve a reliable posture for workloads running on the cloud. The teams curate a list of best practices, in addition to identifying and defining the appropriate operational procedures to optimize their cloud-native applications.

In order to ensure successful cloud migrations, it is important to have a detailed understanding of the capabilities of the cloud resources. Adapting to existing operational approaches and fitting into the cloud platform is not easy. CloudOps requires an organizational-wide change of thinking. It is natural to wonder why it is essential to have a dedicated team to work on cloud IT operations. The essence of innovating on operations is about achieving a smarter cloud. CloudOps comes with advantages as well as challenges. Understanding both will help you to prepare before executing a strategy for CloudOps practices.

Challenges of CloudOps

CloudOps comes with its own set of challenges when it comes to optimizing reliability, agility, and operational management. Let's take a look at some of the main challenges:

- **Complexity**: Companies that subscribe to multiple cloud tools without a proper strategy or giving enough consideration may end up having to maintain the tools. Not having a proper understanding of the complexity makes it daunting for professionals to maintain these solutions.

- **Skills gap**: The skillset in multiple cloud disciplines has been in deficit over the past few years. As cloud migration is evolving, organizations are investing to develop more expertise and narrow the skillset gap to manage cloud workloads.

- **Budget and stakeholder support**: Getting a budget to invest in dedicated teams can be a challenge, especially in the current macro-economic environment. As leaders, it is important to plan for operational and **total cost of ownership** (TCO) savings and benefits for a wise decision staged for growth.

- **Security and threats**: Many cloud vendors offer end-to-end security tools today but you need to be aware of potential vulnerabilities to ensure that the attack surface is zero. The cloud introduces a potential for new attack surfaces that needs to be contained from day one of your modernization journey. Planning through that in the early stages of preparing for cloud migration is crucial.

- **Vendor lock-in**: Working with a cloud provider can lead to vendor lock-in, as we previously discussed. To address this, make sure to weigh your options that meet your growing business requirements.

Keeping in mind the aforementioned challenges and turning them into opportunities while planning an operations strategy can be key to shifting your cloud operations toward agility and overcoming roadblocks.

Advantages of CloudOps

With the right strategy, CloudOps enables organizations to boost the process of building and delivering services that are running on the cloud in many ways. Let's take a look at how organizations can gain benefits from implementing CloudOps practices:

- **Improved reliability**: One of the objectives of the CloudOps philosophy is to have continuous operations and zero downtime. Cloud-based workloads can be updated, deployed, and managed without disruption to the application or service. Implementing CloudOps brings in the essence of automation and redundancy at the cloud-provider level as well as at the application-layer level. This helps to ensure that your applications are up and running despite updates or software patches.

- **Reduced costs**: Companies can reduce the personnel involved in basic resource operations and maintenance, allowing them to avoid associated costs.

- **Improved scalability and flexibility**: CloudOps facilitates standard operating procedures that bring in standardization, instrumentation, and automation. As a result, organizations can achieve faster delivery with quick adjustments and increased efficiencies.

- **Shorter time to market (TTM)**: CloudOps requires organizations to shift from provisioning and managing static infrastructure to dynamic infrastructure on the cloud. This also includes embracing ephemerality and managing a dynamic volume and distribution of services on multiple target environments. CloudOps offers tools to automate tasks such as infrastructure provisioning, creating builds, and running **quality assurance** (**QA**) tests. This helps to reduce time spent on creating and managing manual workflows. As a result, application delivery speed is greatly improved and improves overall service delivery speed.

- **Reduced security risks**: CloudOps facilities use automated implementation processes such as the configuration of cloud resources while provisioning. This helps in automating security checks and establishing clear compliance policies across all teams in a consistent manner. This is particularly essential as your cloud footprint grows and evolves.

In summary, adopting the CloudOps philosophy brings great convenience and value that many enterprises want to make use of. In the next section, we will cover some best practices such as building a CCoE to drive cloud-enabled transformation successfully.

Building a CCoE

When it comes to ensuring cloud transformation success, it is crucial for organizations to have the right skillset and structure in place. A proven way to optimize this strategy is to set up a centralized team that oversees cloud operations, called a CCoE. Some organizations call this internal body a cloud engineering, innovation council, or cloud platform team.

A CCoE is a centralized enterprise governance-functioning team that envisions and establishes a cloud strategy and creates governance principles and best practices for all the teams in the organization to ensure business outcomes with optimal usage of resources. This team typically comprises operations managers, system architects, application developers, network engineers, security analysts, database engineers, and IT managers.

As per Gartner:

"Cloud center of excellence accelerates the uptake of new technologies and optimizes the core capabilities with higher efficiency and lower costs."

When it comes to the core responsibilities of a CCoE, the following are granular-level tasks:

- Championing the cloud transformation
- Planning the logistics for cloud adoption
- Providing guidance on the transition
- Supporting employee training
- Establishing cloud framework baselines and tools for a successful cloud transformation
- Standardizing processes for cloud adoption
- Communicating guidelines and potential risks or issues in a timely manner
- Reviewing process changes for continuous improvement
- Aligning with the business objectives of the organization

The following diagram depicts the various core functions that a CCoE leads to accelerate your organization's cloud computing capabilities and governance:

Figure 12.1 – Cloud Adoption Framework (CAF) on AWS

As organizations witness major IT transformations and a shift in the culture, a new way to manage cloud costs is crucial. There is a need for governance and controlling costs across organizations, which we will cover more in the *Cloud financial management on AWS* section of this chapter. An important aspect to note is that the duties of the team members of a CCoE can be broad as well as deep, depending on the business outcomes of that company, as illustrated in the following diagram:

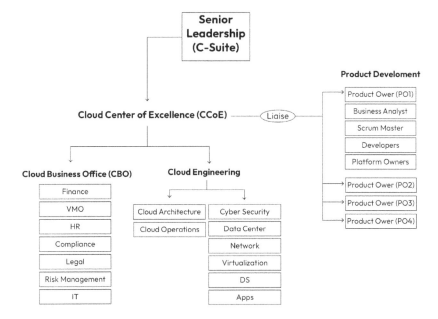

Figure 12.2 – CCoE

Once established, there are two functional subgroups within a CCoE, as shown in the preceding diagram:

- **Cloud Business Office (CBO)**: The CBO team is typically responsible for establishing policies and standards that enable teams to align across **business units (BUs)** and users. This team owns business decisions for aspects such as enterprise architecture, governance, finances, staffing, risk management, change management, vendor management, and communications. The CBO team is led by a cloud transformation leader who acts as chair and is the first-line-of-defense leader when it comes to authority. These are the general goals and responsibilities of a high-functioning CBO team:

  - **Stakeholder alignment**: The CBO team engages with all stakeholders and communicates with them to align decision-making and ownership related to cloud workloads.

  - **Business alignment**: The CBO team engages with business leaders and establishes activities to ensure successful business outcomes and alignment of activities.

  - **Governance model**: The CBO team establishes an organizational structure and a governance framework to manage the evolving needs of the organization and scale the model to meet the needs of future growth.

  - **Best practices**: The CBO team establishes common cloud policies, standards, and templates for the organization through enablement programs such as workshops or training sessions to maintain a culture of learning and development of cloud knowledge across teams.

  - **Strategy and design**: The CBO team develops a cloud strategy and architecture to assess the cloud platform from time to time and identifies baseline activities to set up, configure, and use cloud resources in an optimal way.

  - **Championing the benefits of the cloud**: The CBO team will be responsible for evangelizing the benefits of cloud transformation for the BUs of an organization that hasn't signed up for cloud migration. The team helps to establish and communicate the benefits to encourage full participation and alignment with a sense of ownership.

  In summary, the CBO team has a long list of responsibilities, and diving deeper into them, there are key aspects of making it successful. For additional guidance on creating a CCoE, I recommend looking at AWS's best practices for creating a CBO (https://aws.amazon.com/blogs/enterprise-strategy/creating-the-cloud-business-office/).

- **Cloud engineering team**: This team is responsible for managing and developing the technical capabilities of cloud engineering and infrastructure teams. This team will be considered **subject-matter experts (SMEs)** to advise and guide everyone through reference architectures of applications and workloads to ensure optimal security and performance. This team is led by a cloud engineering leader who chairs activities and provides a vision for scaling motions across the BUs of an organization. These are the goals of a highly functional cloud engineering team:

  - **Migration**: Setting up a strategy to execute migration and modernization activities in an optimal manner.

- **Usage visualization**: Exploring various tools provided by the cloud provider or third-party **software-as-a-service (SaaS)** providers to forecast and manage resourcing needs.

- **Operational efficiency**: Setting up cloud processes and tools to enable desired outcomes of performance, reliability, and efficiency of cloud resources.

- **Cost management**: Working with the finance stakeholder team to get the budget approved and ensuring that BUs do not overspend. The cloud engineering team will also be managing, controlling, and providing guidance with respect to cloud service consumption.

- **Services**: Developing a baseline framework and account structure, including constructs and strategies for a **virtual private cloud (VPC)**, **single sign-on (SSO)**, **disaster recovery (DR)**, and the **Domain Name System (DNS)**.

In summary, defining these lines of responsibility will be helpful for each team member to consistently keep looking for ways to optimize transformation across the organization. It is important to define and map these lines of responsibility to ensure optimal usage of cloud resources across the organization.

What makes a CCoE successful?

Transforming an on-premises operational model to a cloud-native one is one of the core challenges of a CCoE. Removing the complexities involved in this transformation can establish a CCoE as a powerful tool for your organization. Many companies often fail to take crucial aspects of cloud operations—such as governance standards, modernization dependencies, and infrastructure dependencies—into account, but asking the right questions while formulating cloud strategies allows a CCoE to think about these challenges and establish a plan to solve them. This helps make cloud adoption a cost-effective path for your organization. Let's look at some of the important tenets of a CCoE.

Tenet #1 – set up a strong team

The most important aspect of a highly functional CCoE is to set up a strong and multidisciplinary team that is experiment-driven. A team that can learn from failure can iterate quickly. Having a team of cloud architects that understand platforms—AWS, Microsoft Azure, or Google Cloud—can be very beneficial when having difficult discussions with teams while onboarding the cloud.

Tenet #2 – scale, grow, and optimize

The ability to scale, grow, and optimize is the key to making a CCoE successful. Enabling the CloudOps philosophy to drive the cloud transformation agenda and developing strategic initiatives will help align the action plans of individual teams. This will enable them to achieve their business objectives and optimize cloud resources for cost, security, compliance, and utilization.

Tenet #3 – build reusable patterns

With the constant rise in cloud adoption, a CCoE should make a conscious effort to curate reference architectures that can help build an optimal combination of performance, efficiency, high availability, and cost. Evaluating the capabilities of cloud services to use as building blocks for the products that your teams will be building improves the efficiency of the teams. Additionally, frontloading research with a cross-functional team mindset can help in constant innovation and transformation.

Tenet #4 – engage and evangelize

Engaging with your teams across the organization to share information on the latest cloud services and conducting training sessions and workshops gives teams the confidence to experiment with new technologies and optimize cloud workloads. A CCoE should be the guiding force behind this aspect and should establish measures to set up a regular cadence of such engagements.

Incorporating the preceding four tenets should ensure that your company leadership will be able to see the results of your CCoE endeavors more quickly. To better visualize the success that a CCoE can potentially bring with all the right measures, the following diagram compares the before-and-after states of a successful CCoE implementation. Use this as a baseline metric wherever applicable to continuously measure outcomes:

Comparing the before and after of a successful CCoE implementation

Figure 12.3 – CCoE implementation before-and-after states

Throughout the life cycle of your company's cloud journey, the CCoE will liaise between different teams, promoting collaboration, transparency, and alignment of the action plans of individual BUs to the company's general business objectives. In summary, a CCoE's responsibility for shaping the cloud strategy of an entire organization is not an easy feat. Following the guidance in this section and collaborating with the cloud platforms to ensure you have all the required tools to manage, monitor, and mitigate risks will be the key to a successful journey.

Are you well architected on AWS?

Designing applications or technology solutions is a determining factor in how well they can deliver in terms of expectations and requirements. If the foundation is not solid, it can create issues that impede its running state. When you are building applications on AWS, the fundamental guidance is to align with the six pillars of the Well-Architected Framework.

The AWS Well-Architected Framework is based on six pillars: **Operational Excellence**, **Security**, **Reliability**, **Performance Efficiency**, **Cost Optimization**, and **Sustainability**. In this section, let's explore the best practices and design principles for each of these pillars.

Operational Excellence

The **Operational Excellence** pillar provides guidance to run workloads effectively, get insights into the operations of workloads, and improve operational processes, all in support of your business objectives and delivering business value. The four main focus areas to achieve operational excellence are set out here:

1. Knowing your organization's business objectives will help in structuring out your role in it and the expected outcomes. Evaluating customer needs, internal stakeholder requirements, and compliance requirements can help with defining priorities and being able to manage benefits and risks. Understanding the operating model will help focus on efforts to maximize the benefits. Having a well-defined operating model can simplify operations and limit the support overhead of your operating model.

2. Preparing the tools to enable integration, deployment, and delivery of your workloads. Baselining your workload's expected behavior, designing telemetry to understand its current state, and integrating the telemetry back to improve and accelerate deployment activities will help configure your systems for operational readiness in a production environment.

3. Understanding workloads' health, the health of your operations, and risks prior to deploying workloads on production. This helps in gaining visibility and analyzing workload logs. Anticipating operational events—both planned and unplanned—should be part of every organization's incident management playbooks. Defining and responding to alerts related to such events will help you to minimize the business impact. Performing a **root cause analysis** (**RCA**) to prevent the recurrence of such failures should be part of every organization's **key performance indicators** (**KPIs**). I will be providing prescriptive guidance on this in the *Incident management*

section of this chapter. Driving improvements in your business and operational metrics as the business needs change is also important. Evolution is part of the continuous cycle of improving existing operations activities and evaluating the ongoing success of your workloads. Analyzing operational activities and failures and making improvements will ensure operational excellence as your engineering teams grow. Incorporating lessons learned and improvements will help with identifying edge cases and curating best practices across your organization.

When it comes to operational excellence on AWS, there are services available to achieve this. The best practice is to use services such as AWS Control Tower, which can help with expanding the management capability.

Security

The **Security** pillar guides data protection through the design principles in the cloud, as listed here:

- Having a strong identity foundation by implementing the **principle of least privilege** (**PoLP**) enforces the separation of duties within an organization.

- Implementing traceability through monitoring, alerting, and auditing actions that change your environment at any point in time.

- Applying a **defense-in-depth** (**DiD**) approach to incorporate security at layers such as VPC, load balancing, operating system, application, and code.

- Automating security best practices and mechanisms will improve security at scale cost-effectively. Architecting applications with security in mind and defining controls is recommended.

- Protecting data in transit and at rest using mechanisms such as encryption, tokenization, and access control.

- Keeping people away from data using tools will reduce the risk of inappropriate handling of sensitive data.

Preparing for security events and incidents by running game days and simulations with automation will help to accelerate the path to detection and recovery. AWS has established a Shared Responsibility Model that has a well-defined structure for achieving a stronger security posture. AWS offers the **Identity and Access Management** (**IAM**) service, with which you can control user and programmatic access to AWS services and resources on the cloud. With services such as CloudTrail logs, AWS API calls, and CloudWatch monitoring, your organization can achieve continuous monitoring. As we discussed in *Chapter 11, Implementing Security on the Cloud using AWS*, you can capture the required services to protect your resources on AWS.

Reliability

Reliability on the cloud refers to the ability of a workload to operate through its life cycle and perform its intended functionality consistently. When it comes to availability requirements, there are multiple aspects to consider within an application or a service.

On AWS, services are commonly divided into the **data plane** and the **control plane** In general terms, the control plane is responsible for configuring the environment, and the data plane is responsible for delivering real-time service functionality. Examples of data-plane operations are Amazon **Elastic Compute Cloud** (**EC2**) instances, Amazon **Relational Database Service** (**RDS**) databases, and Amazon DynamoDB table read/write operations. Operations such as launching new EC2 instances or RDS databases and adding or changing table metadata in DynamoDB are control-plane operations. High levels of availability are important all across the system, but the design goals should typically index the data plane than the control plane. Avoiding runtime dependency on control operations is generally recommended.

There are a few pointers to keep in mind while designing your systems for high reliability, as outlined here:

- The ability to automatically recover from failure should be a measure of business value. This enables mechanisms to establish monitoring for KPIs and automatic notification of failures. The ability to anticipate and remediate failures ahead of time enables sophisticated automation.

- The ability to anticipate failure scenarios and validate recovery procedures ahead of time. This approach exposes scenarios that you can test and fix, thereby reducing the risk.

- The ability to scale horizontally to make your workloads highly available when a failure is impacting your workloads. Distributing requests across multiple resources ensures that they don't share a common point of failure.

- The ability to satisfy demand without over- or underprovisioning helps protect workloads from resource saturation. Typically for on-premises workloads, when demand exceeds the capacity of that workload, the optimal level is breached, which is the objective of **denial-of-service** (**DoS**) attacks.

- The ability to manage change through automation enables you to track and review changes.

With AWS, developers can implement reliability best practices through AWS **software development kits** (**SDKs**) or the Amazon Builders' Library (`https://aws.amazon.com/builders-library/`), which provides turn-key solutions to operate software in a highly scalable and reliable manner.

Performance Efficiency

This pillar highlights the ability to use computing resources efficiently and maintain efficiency as demands change to meet your business requirements. In order to take a data-driven approach and build a high-performance architecture, it is important to gather data and incorporate the following guidelines:

- Have your cloud vendor implement advanced technologies that can eliminate operational overhead around resource provisioning and management.

- Design workloads that can be deployed in multiple geographic locations to allow low latency and a better experience for your customers at minimal costs.

- Go serverless wherever possible so that you don't have to worry about running and maintaining physical servers for traditional compute activities.

- Experiment continuously to carry out comparative analysis using different types of instances, storage, and configurations

- Explore various cloud services and verify the technical approach that will align best with your business goals.

AWS has a broad range of ecosystems of solution architects, reference architectures, and the **AWS Partner Network** (**APN**) to help you select an architecture and benchmark to optimize your workloads' performance.

Cost Optimization

The **Cost Optimization** pillar advocates for the ability to run systems and deliver business value at an optimal price point. There are a few guiding principles to achieve cost optimization in the cloud, as follows:

- Implementing cloud financial management will help you establish financial success for your company. Dedicate time and resources to build capabilities that will accelerate business-value realization in the cloud.

- Pay only for the resources that your workloads require and depending on the business requirements.

- Measure what you use to know the gains you are making from increased outputs and reduced costs. Analyze IT costs and make your individual workload owners accountable for their spend on resources.

- Stop spending on operations such as racking, stacking, and powering servers, and focus on your customers and business objectives.

When you migrate to the cloud, incorporate financial management on the cloud to achieve business value and financial success. This approach helps in implementing organizational-wide knowledge to build faster and be more agile. We will cover this in more detail in the next section, on cloud financial management on AWS.

Sustainability

This pillar is an important lever to minimize environmental impacts in terms of energy consumption and efficiency. The following are the guiding principles to achieve sustainability while building applications on the cloud:

- Establish sustainability goals such as right-sizing compute and storage resources required per transaction for your workloads. Investing to improve sustainability for existing workloads and giving sustainability goals to BU owners will help in supporting the company-wide sustainability goals of your business or organization. Identify and prioritize areas for potential improvement and model the **return on investment** (**ROI**) of sustainability improvements.

- Maximize utilization of the underlying hardware by right-sizing workloads. To get the most out of underlying hardware on the cloud, it is important to have an understanding of the wide selection of the Amazon EC2 instance types, for instance. Two hosts running at 30% utilization are less efficient than one host at 70% for an estimated baseline on how power consumption should be looked at. Minimizing or eliminating idle resources to reduce the total energy used is the recommended practice. The Instance Scheduler on AWS solution can be used to start and stop Amazon EC2 and Amazon RDS instances.

- Understand key levers to estimate and reduce impact to support upstream improvements for your partners and suppliers. Monitoring and analyzing new and efficient offerings on an iterative basis will help in being flexible to allow for the adoption of new and efficient technologies.

- Use managed services to reduce the amount of infrastructure and operations needed to support your workloads running on the cloud. AWS Fargate for serverless containers, Amazon S3 Lifecycle configurations, or Amazon EC2 Auto Scaling to adjust capacity to meet your demands are a few examples. This helps in sharing common data center components such as power, networking, automation, orchestration, security, compliance, and storage infrastructure.

On AWS, you can use Amazon Monitron, a **machine learning** (**ML**) service to detect and establish behavior patterns in industrial machinery. Using this data, you can reduce the risk of environmental incidents caused by unexpected equipment failures.

Cloud financial management on AWS

As per the 2022 **International Data Corporation** (**IDC**) report, 92% of companies receive important benefits that help companies increase growth, improve business and IT agility, and realize key long-term cost reductions. By now, we have established that economies of scale are one of the key levers for you to start your migration journey to the cloud. But it's not an easy journey, especially for organizations that have complex workloads, and combined with today's unprecedented friction or blockers. It is important to understand and achieve the full potential business value of the cloud. The following diagram showcases a fundamental cloud model on AWS versus a typical charging model on a data center. As per 451 Research (`https://pages.awscloud.com/rs/112-TZM-766/images/451Research-cloud-financial-management-benefits-go-beyond-cost-savings.pdf`), the value of the cloud is not just limited to TCO reduction. You experience a flywheel of greater cloud adoption, higher revenue, and improved profitability:

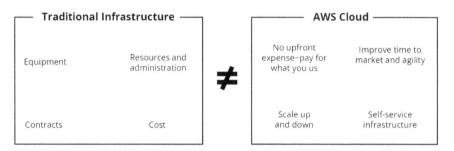

Figure 12.4 – Traditional infrastructure versus AWS Cloud

AWS offers guidance on best practices and advice to financial leaders of organizations, which can help with addressing economic opportunities and making the most of every stage of your journey to the cloud.

Practical recommendations for cloud cost control on AWS

As your cloud adoption journey begins, you will need to implement measures and best practices for cloud financial management. The idea is to incorporate this as a continuous improvement activity. When getting started with cost control, it is important to use the right foundational blocks.

Let's explore a few financial management practical recommendations while operating on AWS:

- **Plan and evaluate**: Knowing where to start to get visibility into your cost and usage information is an essential first step. Whichever approach you plan to build into your cost allocation model, having a clean AWS account structure is a foundational best practice to begin with. Have a designated team or a CCoE design the account structure protocol and have all your teams implement it across your organization for consistency. Having the right level of hierarchical account order and logical groupings will help in maintaining consistency. You can leverage **AWS IAM**, **AWS Control Tower**, and **AWS Organizations** for setting up your account structure and desired policies. If you have a multi-account environment, AWS Control Tower allows you to set up and provision quickly through best-practice blueprints, to ensure that the tasks such as SSO, centralized logging, and administrator activities are preconfigured correctly.

AWS cost allocation tags provide you the capability to label an AWS resource in a unique way. Each tag contains two parts: a key and a value. These tags can be used to organize your AWS resources and can be activated on your cost allocation reports. This makes it easier to categorize and track your AWS costs. You can make use of user-defined tags, which is especially useful if you have multiple business organizations in your company and would like to categorize resources per BU.

- **Monitor cost and usage proactively**: It is important to have an easy way to access spend information with a breakdown of costs across the organization, along with any discounts, deals, or credits that the cloud provider offers. Having real-time visibility on cost and usage information will help in making informed decisions. It gives more context to your engineering, application, and business teams and makes them accountable for their own spending.

  AWS Cost and Usage Reports (**CUR**) provides resource-level cost and usage data that you can review, itemize, and organize for your AWS accounts. You can track your savings plan usage and integrate it with other analytics services such as Amazon Athena for data analysis in a **single pane of glass** (**SPOG**). CUR produces raw data, which is not an ideal way that any business would want to continuously monitor. To address this, you can explore the **Cost and Usage Dashboards Operations Solution (CUDOS) Dashboard**. This is an intuitive set of six dashboards that provide the most comprehensive cost and usage details, with resource-level granularity to enable you to optimize cost and track your usage. The dashboard is built using Amazon Quicksight and lets you build visualizations that can dive deep into details on each chart.

- **Develop a cloud cost optimization strategy**: With the tools in place, developing a solid cost optimization strategy that takes your business goals into consideration will be a crucial factor in assuring a successful cloud journey. Having a dedicated team—ideally, part of your CCoE—is recommended to evaluate cloud spend across your organization. Strategies should include evaluating the cloud spend, identifying the root causes of wasteful spending in the cloud, and choosing the right pricing model. On AWS, you can take charge of your spend by leveraging a variety of pricing models and resources that you can choose from to meet performance and cost-efficiency requirements. **AWS Trusted Advisor** provides insights on potential areas for optimization across areas such as EC2 reserved instance optimization, low utilization of EC2 instances, idle elastic load balancers, underutilized **Elastic Block Store** (**EBS**) volumes, unassociated IP addresses, and idle database instances on Amazon RDS.

  AWS Budgets provides capabilities to set custom budgets and trigger alerts when the usage/spend exceeds a forecasted budgeted amount. **Amazon S3 Analytics** provides automated analysis and visualization of Amazon S3 storage patterns to indicate the ideal storage class to use for better costs. **Amazon S3 Storage Lens** provides visibility into object storage usage and trends on usage and makes recommendations to reduce costs. **Amazon S3 Intelligent-Tiering** provides automatic cost savings on S3 services through data transfers between two access tiers: Frequent Access and Infrequent Access.

AWS Auto Scaling adjusts resource capacity to scale in and out and maintain predictable load performance at the lowest possible costs. **AWS Compute Optimizer** provides recommendations on optimal AWS resources for workloads to reduce costs and improve performance through ML.

The spend in terms of resources also meets the discipline of sustainability goals: reducing waste. As many companies onboard with the cloud financial management function, you will also be able to set and meet sustainability goals.

Cloud financial management (CFM) services on AWS

Cost control on AWS can be done in so many ways, and the following table is an attempt to provide all the services that you can use in a SPOG:

Use case	AWS service	What can you do with it?
Managing and organizing	AWS cost allocation tags; AWS Cost Categories	Design your cost allocation strategy as per your business logic.
Reporting	AWS CUR; AWS Cost Explorer	Track spend with detailed information on resources.
Forecasting and budgeting	AWS Cost Explorer; AWS Budgets	Optimize your resource utilization and keep your spend in check with alert notification mechanisms.
Purchase options	AWS Free Tier; AWS Reserved Instances; AWS Savings Plans; AWS Spot Instances; Amazon DynamoDB on-demand	Choose between free trials and discounts based on your workload patterns.
Elasticity	AWS Auto Scaling; AWS Instance Scheduler; Amazon Redshift Pause and Resume	Scale capacity as per your workload needs.
Right-sizing	AWS Compute Optimizer; Amazon S3 Intelligent-Tiering; AWS Cost Explorer Right-Sizing Recommendations	Align your workloads' actual demand and service allocations proactively.

Table 12.1 – Cloud financial management on AWS

There are popular third-party tools that can help with cost management, such as CloudHealth and Cloudability, and other AWS-compatible tools that can help with overall cost optimization and aligning AWS spend.

SRE on AWS

As the world shifts toward cloud evolution, the reliability of all types of applications running on the cloud has become a critical business imperative. The cloud changes the way we think about managing systems. The focus has shifted from hardware to software and from manual processes to automated tasks. SRE is one such practice that focuses on building sturdy, flexible systems with continuous improvement and automation at its core.

What is SRE?

SRE is the process of automating IT infrastructure tasks to improve the reliability of scalable software systems. Tasks can range from change management, system management, incident response, and application monitoring. The SRE practice aims to improve the stability and quality of the service that is being made available to the end users. You can see a depiction of SRE in the following diagram:

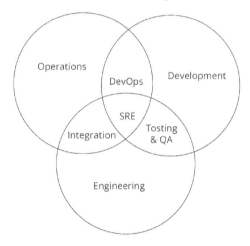

Figure 12.5 – SRE

As shown in *Figure 12.5*, SRE sits right at the crossroads of operations, engineering, and development. SRE is made up of software engineers who automate IT operations that would otherwise be manual processes. The software code automates these tasks in a more scalable and sustainable way to minimize manual interventions that can be error-prone.

Benefits of SRE

SRE culture and practices come with a lot of benefits (as shown in the following diagram), such as cross-team collaboration, decreased downtime, improved customer experience, and better service reliability. Let's dive deeper into these benefits, as follows:

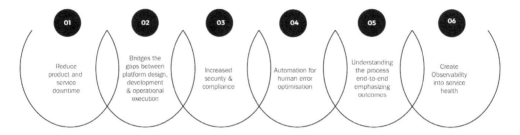

Figure 12.6 – Benefits of SRE

- **Minimizes downtime**: Your teams need to gain SRE knowledge to keep your applications up and running. Continuous testing helps your operations team to reduce downtime. As a result, businesses will value SRE as being more productive and can develop over time to iterate and improve their operations

- **Provides insights into service health**: As your teams incorporate SRE practices, observability will be in place and incident alerting can be automated. (We will talk in detail about some best practices and recommendations for incident management in the next section.) This enables your teams to resolve issues more quickly

- **End-to-end outcomes**: The concept of SRE touches DevOps, connecting your operations and development teams. Streamlining processes reduces complexity across a variety of operations and helps you understand processes from end to end. Your teams will focus on outcomes rather than specific process stages

- **Bridges gaps**: SRE approaches bridge the gap between platform design, development, and operational execution. Your teams will have a better understanding of the platform and the product support effort and will spend more time automating and inventing instead of reactive troubleshooting

- **Improved security and compliance**: SRE's focus on security is increasingly becoming even more popular than the reliability, scalability, efficiency, and performance of the services concerned. Applying security gates on the **continuous integration/continuous deployment** (**CI/CD**) pipeline can help prevent breaches for new deployments

- **Reduces human error through automation**: By now, we have established that automation helps in reducing downtime, particularly human errors as well. SRE makes use of the underlying principle of self-healing processes to proactively identify scenarios that can increase downtime and the cost of fixing errors. Codifying such manual, repetitive, and redundant processes and employing intelligent data-driven methodologies can go a long way toward mitigating them

- **Improved scalability**: A cloud-native SRE approach provides the required framework to facilitate administration, operations, and management to ensure system reliability. This helps in ensuring system reliability for microservices applications at scale for large enterprises whose teams are distributed across several regions

Providing the finest quality services to scale across global applications and infrastructure while ensuring high availability is the core reason why any company would want to add this discipline to their teams. In the next section, we will explore how to implement SRE using AWS services.

Best practices of SRE using AWS

AWS provides specific services for site reliability engineers to optimize time spent on manual tasks related to troubleshooting, observability, and monitoring.

AWS Management and Governance

AWS Management and Governance services are designed to manage and govern your cloud resources at scale. There are a few services as we discussed in the previous chapters, such as AWS Well-Architected Tool, which we already discussed in the prior sections of this chapter. Let's discuss a few services that constitute this category.

AWS Control Tower provides capabilities to set up and govern a secure, multi-account platform called a landing zone in your AWS environment. This is especially useful for many companies with many BUs that implement a multi-account strategy to create a secure and isolated platform for your cloud resources.

Why Control Tower? Building an account topology on the cloud requires a thorough understanding of the best practices and company-wide policies with security and compliance in mind. Control Tower comes with best-practices blueprints and enables governance with guardrails to choose from a pre-packaged list. With all the well-architected best practices and baseline setup guidance for multi-accounts, you can create a landing zone in a few clicks and quickly provision a new AWS environment with hardened security. The following diagram illustrates how AWS Control Tower orchestrates multiple AWS services such as IAM, AWS Organizations, and Service Catalog on your behalf and automates the creation of a landing zone:

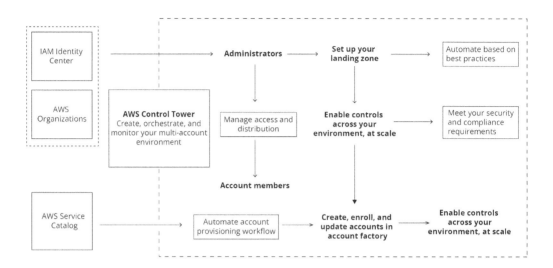

Figure 12.7 – AWS Control Tower

You can create guardrails to manage the ongoing governance of your AWS environment, which helps in detecting the non-conformance of provisioned resources proactively. Control Tower offers blueprints such as creating a multi-account environment using AWS Organizations and providing identity management within AWS IAM Identity Center (successor to **AWS Single Sign-On**, or **AWS SSO**), as well as providing federated access to accounts and enabling cross-account security audits using AWS IAM Identity Center.

AWS License Manager provides a simplified experience for you to effectively govern and manage software licenses such as Windows, SAP, Oracle, IBM, and SQL Server, which typically require a dedicated physical server. You get the combination of enjoying the flexibility of **Bring Your Own License** (**BYOL**) and managing them cost-effectively on Amazon EC2 dedicated hosts. License Manager can help you track your license usage and covers the following use cases:

- Streamlining license management where AWS licenses can be tracked and managed.

- Simplifying the BYOL experience where you can set rules to manage, discover, and report software BYOL license usage.

- Controlling AWS Marketplace license entitlements with automation on activating software entitlements and workloads across AWS accounts for end users

- Managing user-based license subscriptions where you can subscribe, manage, and track user-based licenses. An example is Microsoft Visual Studio on an Amazon EC2 instance.

License Manager integrates with AWS Systems Manager to help with discovering any software installed on your AWS resources.

Incident management

As companies look to grow their digital footprint on the cloud, it is important to have a strategy to safeguard themselves from incidents that can cause application downtime and business disruption. An incident can be anything that impacts your applications from running as expected. The root cause could arise from issues with software, infrastructure, or anything in between.

> **Note**
> Incident management on the cloud is an alignment of operations, resources, and services to manage incidents, maintain an "always-on" service level, keep businesses running, and deliver seamless customer experiences.

When it comes to the best practices for successful incident management on the cloud, the following suggestions are recommended:

- **Detect incidents early**: Detecting critical incidents ahead of time can help with faster resolution and reduce the potential for failures.

- **Streamline operations with real-time collaboration**: Stakeholder teams responsible for triaging and resolving the issue need to have a platform for collaborative and effective communication.

- **Keep the customer informed**: This is the most crucial during the life cycle of the incident, where transparent communication to the impacted consumer will always go a long way in earning their trust.

- **Automate wherever possible**: Automating tasks such as incident ticket assignments and alerting the right members of the team can save a lot of time.

- **Conduct a post-mortem analysis**: Doing a **root cause analysis (RCA)** is a process to identify the nature of the problem as well as its solution. The famous *5 whys* is a problem-solving method to explore the cause-and-effect relationships of a particular problem. It is a structured method of diving deep until you reach the root cause.

In addition to the aforementioned points, maintaining a repository of incidents containing important and relevant data about incidents within a persistent store is recommended. This can be used to generate incident reports for further analysis.

AWS Incident Manager

AWS Incident Manager supports the following four stages of the incident life cycle on the cloud:

1. **Alerting and engagement**: This phase gets triggered when an incident is detected. AWS Incident Manager provides you with options to set up monitoring your applications through CloudWatch metrics and configuring alerting. You can define an incident response plan to work on incidents and use templates to launch escalation or engagements.

2. **Triaging**: In this phase, you can prioritize incidents and define a rating for each of the priorities (1: Critical impact; 2: High impact; 3: Medium impact; 4: Low impact; 5:No impact).

3. **Investigation and mitigation**: In this phase, you can make sure that runbooks, timelines, and metrics are in place. Incident Manager integrates with System Manager and builds runbooks. Timelines help with actions to take, and metrics are automatically populated. Incident Manager also provides a chatbot interface to communicate with the team during the investigation.

4. **Post-incident analysis**: This helps with reflecting on the analysis and formulating processes to improve the response and customer experience. Improvements such as changes to the application, incident response plan, runbooks, or alerting can be incorporated during this phase.

The following diagram illustrates these four phases of the incident life cycle:

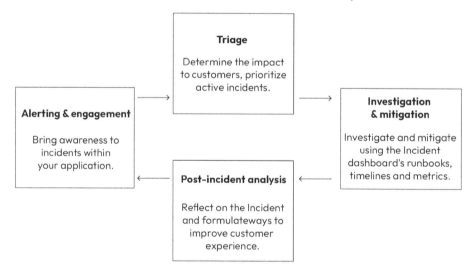

Figure 12.8 – Incident life cycle

AWS Incident Manager integrates with ServiceNow, Jira, and PagerDuty to make it easier for your teams to seamlessly connect and collaborate. One of the benefits of using such a service on AWS while operating on it is that the security model remains the same as if you were using any other service. The APIs provide capabilities of compliance, resiliency, and IAM integration, which makes it seamless.

AWS Health Aware

AWS Health Aware (`https://aws.amazon.com/blogs/mt/aws-health-aware-customize-aws-health-alerts-for-organizational-and-personal-aws-accounts/`) is a solution that any enterprise customer on AWS can use to customize AWS Health API alerts on the cloud. It is a serverless turn-key solution to automate proactive alerting to provide real-time alerts to your preferred communication channels. With extensive customizations and integrations, this tool provides many advantages, including aggregated Health API alerts, along with prescriptive guidance to mitigate issues on the cloud during infrastructure events.

Summary

In this chapter, we learned the key aspects of designing and operating your applications at scale on the cloud. We explored the AWS Well-Architected Framework and its six pillars: Operational Excellence, Security, Reliability, Performance Efficiency, Cost Optimization, and Sustainability. We also discussed practical recommendations to set your business up for success in terms of organizing, reporting, and managing costs. Key concepts of operating efficiently on the cloud such as SRE and incident management were discussed in detail. In the next chapter, we will recap and look at some of the upcoming trends and technologies.

Further reading

- AWS SRE guide: `https://aws.amazon.com/what-is/sre/`
- AWS Health Aware: `https://aws.amazon.com/blogs/mt/aws-health-aware-customize-aws-health-alerts-for-organizational-and-personal-aws-accounts/`
- AWS Well-Architected Framework, Lenses, and Guidance: `https://aws.amazon.com/architecture/well-architected/`
- AWS cloud financial management: `https://aws.amazon.com/aws-cost-management/`
- AWS CUDOS Dashboard: `https://aws.amazon.com/blogs/awsmarketplace/using-cudos-dashboard-visualizations-aws-marketplace-spend-visibility-optimization/`
- Incident management on the cloud guide: `https://docs.aws.amazon.com/managedservices/latest/userguide/what-is-incident-mgmt.html`
- AWS management and governance: `https://aws.amazon.com/products/management-and-governance/`

13

Wrapping Up and Looking Ahead

We've learned how AWS can be instrumental in driving digital transformation across organizations. While cloud migration is the first step to transforming your legacy systems, you will need strategies, technologies, and the right **cloud service provider** (**CSP**) who can support your business needs. Modernizing your systems, containerizing your applications, and using serverless technologies that are fully managed to provide you with a platform that is highly secure, scalable, flexible, and resilient yet also cost-effective will help you embark on large-scale modernization efforts. This will enable you to improve the speed and economics of IT service delivery to boost innovation and provide your customer with new experiences.

Before we close the book, I would like to provide a few time-tested guides and patterns to help with your modernization and optimization projects.

In this chapter, we're going to cover the following main topics:

- Modernization recap—Breaking down your options:

  - For builders

  - For decision-makers

- Emerging trends and technologies:

  - Web3

  - Blockchain

  - Graviton

Modernization recap – Breaking down your options

We learned about migration, which is the first step for any modernization efforts, in *Chapter 3, Preparing for Cloud Migration*, and discussed how a well-planned migration strategy can reduce the complexity of the migration process and set the stage for your legacy systems to be migrated successfully. The 6 Rs of cloud migration—which stand for *relocate, re-host, re-platform, refactor, repurchase*, and *retire*—include the standard patterns for cloud migration whereby you choose the right option depending on the state of your workload. The AWS **Cloud Adoption Framework (CAF)** helps you evaluate cloud readiness across eight dimensions, as shown in the following diagram:

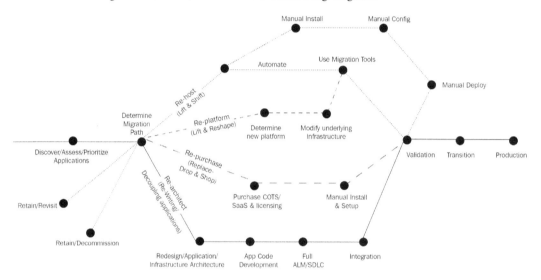

Figure 13.1 – AWS CAF

The modernization journey typically involves two key personas—builders and decision-makers. The following two sections outline the breakdown of strategies with a focus on these personas.

For decision-makers

Operating in the cloud enhances operational efficiency, reduces costs, improves security, and increases agility. The modernization experience for organizations that have a meaningful portion of IT assets is often dreadful if they don't have structured planning. Removing modernization risks is crucial to help ease the transition, and there is no one-size-fits-all solution, but time-tested strategies can often come in handy. To set your organization up for success, start by building a comprehensive strategy and develop an action plan taking what we have learned from *Chapters 3, 4*, and *5* into consideration.

Choosing the right CSP

A CSP is one of the most crucial aspects to ensure long-term success. Key areas for consideration are set out here:

- The kind of security measures the CSP provides
- Whether the CSP is beholden to compliance standards such as the **General Data Protection Regulation (GDPR)**, **Systems and Organizations Controls 2 (SOC 2)**, the **Payment Card Industry Data Security Standard (PCI DSS)**, and the **Health Insurance Portability and Accountability Act (HIPAA)** for the CSP's service roadmap to grow and innovate
- The data governance policies of the CSP
- Service dependencies
- **Service-level agreements (SLAs)** and any terms and conditions offered by the CSP, along with any business terms
- Legal agreements, which should be reviewed and carefully considered

Additionally, understanding the CSP's **disaster recovery (DR)** provisions, migration support, and vendor lock-in constraints are some of the soft and hard factors to take into account for your assessment process. For a comprehensive checklist, review *Chapter 3*.

Ensuring visibility and cost reduction

Many migrations start with a **lift-and-shift** strategy. There is a compounding effect of increasing the **total cost of ownership (TCO)** if you don't ensure ongoing optimization through modernization efforts. Take advantage of services such as AWS Organizations, AWS Budgets, and AWS Trusted Advisor to have full visibility of costs across the organization and to avoid any surprises in costs. Use monitoring tools such as Amazon CloudWatch and assign metadata to your AWS resources in the form of tags. **Tagging** can help you understand and control your AWS costs. You can leverage AWS solutions such as Instance Scheduler to start and stop your **Amazon Elastic Compute Cloud (Amazon EC2)** and **Amazon Relational Database Service (Amazon RDS)** instances. This helps to reduce operational costs and only uses capacity when needed.

Including prioritization

While migrating your applications to the cloud, rank your applications by importance to your business. Additionally, take usage patterns, internal or external dependencies, underlying architecture complexity, and technologies used into consideration. Iterate through these applications in phases with a timeline and a roadmap to provide visibility and for modernization tracking.

Ensuring ongoing knowledge transfer and enablement

If your organization is using an AWS Partner from the **AWS Partner Network** (**APN**), find the right partner that aligns with your organizational culture and that can also document implementations. In parallel, make sure your own teams are upskilled to take advantage of the benefits of the cloud. Structured training such as brown bag sessions, immersion days, and AWS certifications can help boost the confidence of technology teams.

Planning for multiple staging environments

Build multiple staging environments in the cloud and make sure to follow software coding best practices such as version control and parameterized stack templates while using **Infrastructure as Code** (**IaC**). This helps with consistency and reliability and becomes a catalyst for the successful outcome of your migration efforts.

Implementing business testing and ensuring high availability

Spotting potential issues or gaps at the earliest possible stage will help you reduce any downtime of your application. Secure the time of key business stakeholders and have them preview the end-to-end application experience.

As a top-down approach, start by building the culture as a foundational element and build your team of cloud experts, such as developers, database administrators, network engineers, security experts, and finance experts. This institutionalizes best practices and frameworks while your organization moves to the cloud.

In the next section, let's look at some of the key factors to take into consideration if you belong to the builder persona.

For builders

For developers or builders on the cloud, we discussed in *Chapters 5* to *10* how to establish a holistic approach for transforming legacy systems through the following areas of modernization.

Application modernization

As legacy systems expand due to business changes, it becomes difficult to add features and functionalities. Maintaining an ever-expanding code base is often a daunting task. Testing even minor changes requires regressive lengthy procedures, and the development of new features as a result becomes a slow process.

Microservices architecture

Companies of any size can boost agility and flexibility by redesigning their legacy applications into microservices architecture with loosely coupled components. You can implement new features, detect bugs, deploy fixes at the level of a single service, and release features more rapidly.

Asynchronous communication between services aids loose coupling and creates clear transaction boundaries to support the independent operation of services. Microservices architecture enables clear ownership of the design, development, deployment, and operations of life-cycle activities.

We learned how you can use **AWS Step Functions**, a workflow service used to orchestrate and automate business processes, AWS services, and build serverless applications. For API management, you can use **Amazon API Gateway**, a fully managed service that is used to create, publish, maintain, monitor, and secure APIs at any scale.

Containers and serverless

Containers are portable and lightweight and can be run easily at any scale. Containers are an integral part of breaking down the traditional monolithic application architectures, and they enable a transition to microservices for easier scale. AWS offers **Amazon Elastic Container Service** (**Amazon ECS**), a fully managed container orchestration service for running containerized applications in a secure, reliable, and scalable manner. **Amazon Elastic Kubernetes Service** (**Amazon EKS**) is a fully managed Kubernetes service for running containerized applications using Kubernetes in a secure, scalable, and reliable manner.

Serverless and fully managed services

Along with the previously mentioned container orchestration services, we learned in *Chapter 7* about **AWS Fargate**, which removes the need to provision and manage underlying servers—you can instead just pay per usage.

Organizations look for ways to reduce operational complexity, and serverless technologies enable them to do so. As system administrators, you do not have to worry about provisioning and managing servers, spending time on OS patches, or maintaining unused resources that were originally provisioned to meet peak usage demands.

With **AWS Lambda**, you can write custom implementations and not have to manage any underlying infrastructure or worry about scaling and maintaining event integrations.

AWS Lambda is a serverless computing service to enable the implementation of custom code without requiring any provisioning or managing servers. You can create workload-aware cluster scaling logic, maintain event integrations, or manage runtimes.

Database modernization

Legacy databases come with a number of challenges, which include a lack of scalability, high costs, overhead complexity of database administration, and the complex process of hardware provisioning.

In *Chapter 8*, we learned about purpose-built databases such as **Amazon DynamoDB**, which is a key-value and document database that is a fully managed service with security, built-in backup and restore, and in-memory caching.

Additionally, **Amazon Aurora** is a relational database that is MySQL- and PostgreSQL-compatible, highly performant and available, and ideal for traditional enterprise databases. Aurora comes with speed, security, availability, and reliability at one-tenth of the cost of commercial databases.

Amazon RDS is a relational database in the cloud that provides six database engines, including PostgreSQL, MySQL, MariaDB, Oracle Database, Aurora, and SQL Server. RDS is optimized for memory, performance, and I/O to achieve the fast performance, **high availability** (**HA**), security, and compatibility that enterprise applications need.

AWS Database Migration Service (**AWS DMS**) allows you to easily migrate or replicate your existing databases to Amazon RDS.

Amazon Redshift gives you the ability to query and combine exabytes of structured and semi-structured data across your data warehouse, operational database, or data lakes using standard SQL. It has the capability of saving the results of queries into **Amazon Simple Storage Service** (**Amazon S3**) data lake-supported formats such as Apache Parquet. This helps in performing additional analytics through analytics services such as Amazon **Elastic MapReduce** (**EMR**), Athena, and SageMaker.

Data and storage modernization

Storage solutions can be expensive, might not scale well to meet growing demands, and can have durability and availability issues. There is a complete range of services (object storage, file storage, and block storage) from AWS to store, access, govern, and analyze your data to reduce costs, improve agility, and accelerate innovation. Data analysts can elect to back up and utilize the concept of data lakes to build the foundation of modern applications.

A data lake allows you to store both structured and unstructured data at any scale in a centralized repository. Once you store your data, you can use tools to transform the data and run different types of analytics to make decisions using dashboards and visualizations to analyze the data.

Setting up and managing data lakes is often a manual and time-consuming task. Many enterprises run their data lakes depending on various third-party providers of tools for data transformation and data visualization. AWS makes it easy by providing all the required tools and processes to automate such tasks with **AWS Lake Formation**. You can automate the tasks so that you can build and secure your data lake in days instead of months using Lake Formation.

For data storage, **Amazon S3** has gained popularity due to its unmatched durability (11 9s), cost-effectiveness, and availability (99.99%). **Amazon HealthLake** is a fully managed HIPAA-eligible service that enables healthcare customers to aggregate health information from different sources and formats into a structured, centralized AWS data lake. You can use analytics and **machine learning** (**ML**) to extract insights from the data.

There is often a need for data analysts to process data before moving it to data lakes through actions such as combining, moving, or replicating data across multiple data stores. **AWS Glue** provides data integration capabilities that make it easy to discover, prepare, and combine data for analytics, ML, and application development.

Operations on the cloud

Enterprises running legacy systems have the longest resolution times for operational events and extended software release cycles. DevOps practices that include monitoring, logging, and auditing into operations can improve operational agility. Here are some of the processes that we discussed in *Chapters 9, 10*, and *11*:

- **Continuous integration and continuous deployment and delivery (CI/CD)**: CI/CD automation augments well with modern applications. Release processes involve things such as building the code, system testing, integration testing, moving artifacts to staging, and pushing them into production. By automating such release processes, you can deliver the maximum value, and AWS services such as AWS CodeBuild, AWS CodePipeline, and AWS CodeDeploy can help with this.

- **IaC**: You can achieve the maximum benefits of CI/CD through modeling applications using IaC. Using IaC in your application development life cycle can help in faster provisioning and reducing configuration errors. With AWS services such as **AWS CloudFormation**, you can provision resources quickly and consistently. The **AWS Serverless Application Model** (**AWS SAM**) is an open source framework to build serverless applications. The **AWS Cloud Development Kit** (**AWS CDK**) is an open source software development framework to define your cloud application resources using programming languages.

- **Monitoring and logging**: Operational teams need to monitor the behavior of their applications and analyze user-level metrics such as latency, availability, and response times to improve the overall experience. AWS offers a number of services for monitoring and logging. **Amazon CloudWatch** lets you monitor and observe your systems and applications. This is a useful tool for builder personas such as DevOps engineers, developers, and Site Reliability engineers (**SREs**). CloudWatch provides data and actionable insights to monitor applications, respond to system-wide performance changes, and help optimize resource utilization. Logs from multiple sources can be centralized with CloudWatch. This is especially useful during triaging and searching for specific error codes or patterns. **AWS X-Ray** helps developers analyze and debug production applications. X-Ray is most useful for applications built using microservices architecture where the root cause of performance issues and errors may be difficult to trace.

Security on the cloud

Protecting cloud data, applications and infrastructure from threats is a shared responsibility model between the cloud provider and you as a customer. We discussed in *Chapter 10* that the cloud provider will be responsible for the security of the underlying cloud infrastructure, and you are responsible for securing workloads that you run on the cloud. This means that there is a significant area of responsibility that you need to take ownership of.

In this section, we will review the top best practices and the important elements of an AWS security strategy that will ensure your data, applications, and resources are protected. AWS offers several services and features that can provide controls to help you meet your security objectives. However, you cannot protect your resources to the fullest extent without complete awareness of which resources are in use and who needs to use them. Defining your cloud security strategy is an important first step in identifying your business risks and threats. Planning specific AWS services for a wide range of applications involves a comprehensive understanding of the industry-standard best practices across your organization.

Your strategy should consider incorporating the following aspects:

- **Principle of least privilege** (**PoLP**) across your resources on the cloud
- **Multi-factor authentication** (**MFA**) and enabling **single sign-on** (**SSO**)
- Rotating access keys regularly
- Threat and incident response planning
- Visibility across your cloud accounts
- DevSecOps to integrate with your development workflows
- Security automation wherever possible
- **Defense-in-depth** (**DiD**) security layering
- Cloud-native security solutions and platforms to ensure the protection of resources throughout the development process
- Detection, monitoring, and alerting
- Data protection using encryption
- Regular updates and frequent patches to keep cloud resources up to date
- Regulatory compliance for industries such as healthcare, financial services, and government

You can leverage the AWS Partner ecosystem and AWS Marketplace to extend the benefits of using security-focused solutions for specific workloads and use cases. The readily available solutions help you to easily find, buy, and deploy **Software as a Service** (**SaaS**) products to help secure your data in ways not possible on-premises.

Network on the cloud

The network connectivity on the cloud supports your workloads to be highly scalable, performant, and available with broad global coverage. We learned in *Chapter 11* about how the cloud network varies from traditional networks backed up by on-premises servers. The CSPs host the IT infrastructure in a public or private cloud platform to provide the required resources in an on-demand manner using an internet connection. We learned about the networking fundamentals and constructs on the cloud such as **Amazon Virtual Private Cloud** (**Amazon VPC**) to customize and control your networking environment. Let's look at some best practices while using a VPC on your AWS accounts:

- Use multiple A**vailability Zones** (**AZs**) for HA.

- Use security groups to restrict and control the flow of traffic in and out of a VPC.

- Use **Identity and Access Management** (**IAM**) policies as an administration tool to manage access levels of resources.

- Use **Amazon VPC Flow Logs** and send information to CloudWatch to capture operational data in the form of logs, metrics, and events.

- Configure the right design for your VPC by selecting a **Classless Inter-Domain Routing** (**CIDR**) block with a big enough address range for future growth.

- Separate your VPC environments for development, QA, staging, and production environments.

- Secure your Amazon VPC by using an **intrusion detection system** (**IDS**) and **intrusion prevention system** (**IPS**) to protect it from attacks or unauthorized access.

Every network design is different, and using the AWS Well-Architected Framework can help guide you to build resilient and fault-tolerant applications.

Emerging trends and technologies

Public cloud provider usage is shifting among enterprises. The ongoing mass adoption is a key driver for transformative technologies such as AI, the **Internet of Things** (**IoT**), and hybrid working. **Virtual reality** and **augmented reality** (**VR/AR**), the metaverse, and quantum computing are expectedly becoming enablers.

Compute-intensive processors – AWS Graviton

AWS Graviton processors are designed by AWS to deliver cost, energy, and resource efficiency for workloads running on Amazon EC2, and many managed services such as Amazon Aurora, AWS Lambda, AWS Fargate, Amazon ElastiCache, and Amazon EMR support workloads running on Graviton. With a significant leap in performance and capabilities, AWS Graviton processors offer up to 40% better price and performance over current-generation x86-based instances. The latest processor, Graviton3, uses up to 60% less energy relative to general-purpose EC2 instances without compromising on performance. Graviton delivers enhanced security of always-on memory encryption, dedicated caches for every vCPU, and support for pointer authentication.

Graviton3 supports a broad range of compute workloads. The C7g instances powered by AWS Graviton3 processors provide up to 25% better compute performance and make it suitable for a wide range of compute-intensive applications, from web servers, load balancers, and batch processing to **electronic design automation** (EDA), **high-performance computing** (HPC), gaming, video encoding, scientific modeling, distributed analytics, ML interference, and ad serving.

Hybrid cloud – AWS Outposts, AWS Local Zones, AWS Wavelength, and Amazon ECS/EKS Anywhere

From start-ups to the largest enterprises seeking to run their applications on the cloud rapidly, these companies are also looking to bring AWS services closer to where they're needed, such as on-premises. There could be requirements to work with local datasets or even share the data across on-premises applications but with minimal latency. Aligning with data residency requirements is a must. For such scenarios, data entering the cloud is not ideal.

AWS offers a hybrid cloud by extending its infrastructure and services to customers wherever they need it. The following information comes is as per the AWS documentation:

AWS Regions and AZs (`https://aws.amazon.com/about-aws/global-infrastructure/regions_az/?p=ngi&loc=2`): A region is a physical geographic location that has a cluster of AWS data centers within it. These data centers are known as AZs that are grouped in a minimum of three per region to provide redundancy.

AWS Local Zones(`https://aws.amazon.com/about-aws/global-infrastructure/localzones/?p=ngi&loc=3`): A Local Zone is an extension of an AWS Region into a metropolitan area. AWS Local Zones is suitable for applications that require low latency. Examples of such applications include systems built for live event streaming, gaming, smart grid, and intelligent transport systems.

AWS Outposts (`https://aws.amazon.com/outposts/`) provides capabilities to put native AWS cloud infrastructure and services into any data center, colocation site, or on-premise facility. The AWS cloud, its services, and APIs can be extended to locations that can provide local computational capabilities. You can process data locally and avoid bandwidth limitations.

AWS Wavelength (`https://aws.amazon.com/wavelength/`) provides capabilities to have AWS compute and storage services at the edge of the 5G mobile phone network. This enables next-generation mobile app developers to build mobile applications with sub-millisecond latency.

Amazon ECS/EKS Anywhere (`https://aws.amazon.com/ecs/anywhere/` and `https://aws.amazon.com/eks/eks-anywhere/`) provides capabilities to manage deployments in Regions, on Outposts and Wavelength, and in AWS Local Zones and integrate the operational processes to manage containerized workloads on AWS and non-AWS platforms. This enables you to deploy native container tasks in any environment. This also makes it favorable to support low-latency applications and manage a single control plane running in the cloud.

Web3

Web3 is the third generation of the **World Wide Web** (**WWW**).

The following diagram shows the history of how the web evolved from Web1 to Web3. Web1 was created in 1989 to cater to static web pages. Web2 in 2004 was introduced to support web and mobile applications, with Web3 was introduced in 2019 to support **decentralized applications** (**DApps**):

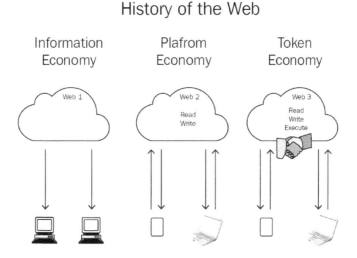

Figure 13.2 – Web evolution

Web3 is emerging as the next fundamental shift of the emerging technology ecosystem that could revolutionize the digital business model. Web3 is built on the concept of blockchain technology and a **peer-to-peer** (**P2P**) network with no central authority. Blockchain technology establishes a verifiable and transparent way to ensure and retain control of the user's personal data.

Primarily run on Ethereum (a type of cryptocurrency), Web3 has the following key characteristics:

- A decentralized web puts the power back in the hands of the users and enables end users or devices to connect without the need for centralized infrastructure. Users own their data and have control over how they want to share the data.

- Edge computing infrastructure plays a fundamental role in Web3's decentralized nature. With the use of edge infrastructure, Web3 is uniquely positioned to run in a globally distributed environment.

- The Semantic Web is Web3's ability to transform internet data into machine-readable data. This enables cross-chain data sharing across enterprises.

- Connectivity across the web for a next-level digital experience is possible through the Web3 ecosystem across various data sources and devices such as IoT sensor-based, thereby delimiting access to smartphones and computers.

- AI in Web3 can play a significant role in creating more intuitive interfaces for web applications. For example, AI can help build user-friendly interfaces by taking each user's personal preferences into consideration. This helps to enhance the usability of the application.

- Transparency in Web3 enables the user community to make informed decisions with real-life impact.

Let's take a look at the existing and possible future use cases of Web3.

Use cases of Web3

The following list presents some existing and possible use cases of Web3:

- **Next-gen DApps**: DApps are the next evolution of the web. They are inherently decentralized and allow data to cryptographically flow between the intended users. **Decentralized finance (DeFi)** apps can be created to give users more control over their money through personal wallets. Cross-chain apps that are an emerging concept for DApps involving smart contracts between different blockchain networks can be supported.

- **DeFi**: Supports multichain DeFi solutions that are highly efficient and scalable.

- **Advanced gaming**: Supports pay-to-earn, play-to-own, and crypto-based games.

- **Social media**: Supports social media DApps and wallet-based and private key-based applications.

- **NFT**: Supports transactions that contain immutable **non-fungible token** (**NFT**) records and tokenization.

- **Real estate**: Supports NFT-backed properties, digital proof of ownership, and a 3D real-estate marketplace.

- **Remote workplaces**: Supports interactive virtual meetings, avatars, and 3D-enabled workplaces.

There are many possibilities when it comes to AWS offering capabilities to put control in users' hands and empower flexibility to developer communities. With AWS services such as AWS Lambda, you can build DApps compatible with Web3. You can run Web3-compatible JavaScript code and integrate it with the Node.js SDK. You can leverage the benefits of Lambda, such as automatic scaling and cost reductions, and deliver DApps with a higher level of data privacy and security.

AI

AI refers to any model that is capable of performing tasks requiring human intelligence. AI engineering is playing a critical role in turning complex AI engineering pipelines into self-adaptable systems in production. As per Gartner's *Top Strategic Technology Trends 2023* report (`https://emtemp.gcom.cloud/ngw/globalassets/en/publications/documents/2023-gartner-top-strategic-technology-trends-ebook.pdf`), enterprises that adopt AI engineering practices to build their adaptive AI systems will outperform their peers in operationalizing AI models by at least 25%. AI engineering models' Adaptive AI makes it easier for enterprises to maximize business value through the models and AI engineering practices and streamlines design patterns and event-stream capabilities, as illustrated in the following diagram:

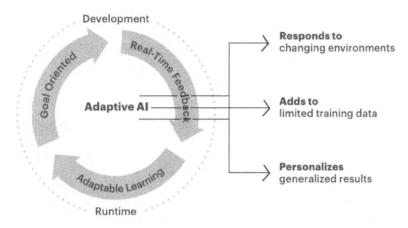

Figure 13.3 – Adaptive AI

As AI continues to advance, AI models such as ChatGPT are making breakthroughs, gaining popularity, and influencing many businesses to think about all the things that AI can achieve in the next 5 to 10 years. This will potentially change how industries will reorient their business models and become pioneers in changing the way people work, learn, travel, and communicate with each other. Given the rapid pace of change, these AI models will improve decision-making in industries such as healthcare, finance, and many more. The advancement of data analytics to feed ML models will accelerate our ability to perform tasks that otherwise would have taken years.

Using AI to improve data quality will exponentially push many businesses to take up initiatives and develop new business models in industries such as agriculture, healthcare, and finance and get tremendous value without investing in big effort or money. AI adoption is predicted to skyrocket, and new-age applications will continue to break the boundaries between reality and science fiction. Business leaders can improve their decision-making by encouraging and investing in research on these new technologies.

Summary

In this chapter, we reviewed the lessons learned through all the chapters and aspects of digital transformation such as application modernization, transforming security, and networking on the cloud. We also covered emerging technologies such as Web3, compute-intensive processors, and the hybrid cloud.

Thank you, and take action!

All of you are given two great gifts: your mind and your time. With each opportunity that presents you to take action, you—and only you—have the power to determine the changes you want to bring in. Invest in learning, share this knowledge, and be prepared to take the necessary actions to make improvements.

Further reading

Here are some additional resources to learn from as you continue your reading journey:

- *Architecting Cloud Computing Solutions* by Kevin L. Jackson , Scott Goessling: `https://www.packtpub.com/product/architecting-cloud-computing-solutions/9781788472425`

- *Cloud Native Architectures: Design high-availability and cost-effective applications for the cloud by* Erik Farr, Kamal Arora, and Tom Laszewski: `https://www.amazon.com/Cloud-Native-Architectures-high-availability-cost-effective/dp/1787280543`

Index

Other Books You May Enjoy

If you enjoyed this book, you may be interested in these other books by Packt:

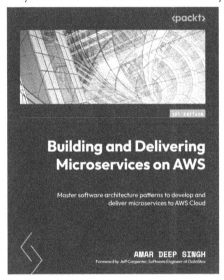

Building and Delivering Microservices on AWS

Amar Deep Singh

ISBN: 978-1-80323-820-3

- Understand the basics of architecture patterns and microservice development
- Get to grips with the continuous integration and continuous delivery of microservices
- Delve into automated infrastructure provisioning with CloudFormation and Terraform
- Explore CodeCommit, CodeBuild, CodeDeploy, and CodePipeline services
- Get familiarized with automated code reviews and profiling using CodeGuru
- Grasp AWS Lambda function basics and automated deployment using CodePipeline
- Understand Docker basics and automated deployment to ECS and EKS
- Explore the CodePipeline integration with Jenkins Pipeline and on premises deployment

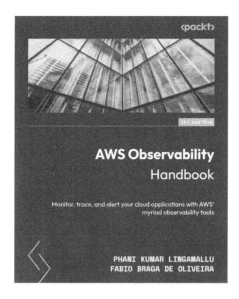

AWS Observability Handbook

Phani Kumar Lingamallu, Fabio Braga de Oliveira

ISBN: 978-1-80461-671-0

- Capture metrics from an EC2 instance and visualize them on a dashboard
- Conduct distributed tracing using AWS X-Ray
- Derive operational metrics and set up alerting using CloudWatch
- Achieve observability of containerized applications in ECS and EKS
- Explore the practical implementation of observability for AWS Lambda
- Observe your applications using Amazon managed Prometheus, Grafana, and OpenSearch services
- Gain insights into operational data using ML services on AWS
- Understand the role of observability in the cloud adoption framework

Packt is searching for authors like you

If you're interested in becoming an author for Packt, please visit `authors.packtpub.com` and apply today. We have worked with thousands of developers and tech professionals, just like you, to help them share their insight with the global tech community. You can make a general application, apply for a specific hot topic that we are recruiting an author for, or submit your own idea.

Share Your Thoughts

Now you've finished *Optimizing Your Modernization Journey with AWS*, we'd love to hear your thoughts! Scan the QR code below to go straight to the Amazon review page for this book and share your feedback or leave a review on the site that you purchased it from.

`https://packt.link/r/1803234547`

Your review is important to us and the tech community and will help us make sure we're delivering excellent quality content.

Download a free PDF copy of this book

Thanks for purchasing this book!

Do you like to read on the go but are unable to carry your print books everywhere?

Is your eBook purchase not compatible with the device of your choice?

Don't worry, now with every Packt book you get a DRM-free PDF version of that book at no cost.

Read anywhere, any place, on any device. Search, copy, and paste code from your favorite technical books directly into your application.

The perks don't stop there, you can get exclusive access to discounts, newsletters, and great free content in your inbox daily

Follow these simple steps to get the benefits:

1. Scan the QR code or visit the link below

https://packt.link/free-ebook/9781803234540

2. Submit your proof of purchase
3. That's it! We'll send your free PDF and other benefits to your email directly

www.ingramcontent.com/pod-product-compliance
Lightning Source LLC
Chambersburg PA
CBHW081503050326

40690CB00015B/2912